SPRINGER PROTOCOLS HANDBOOKS

For further volumes:
http://www.springer.com/series/8623

Springer Protocols Handbooks collects a diverse range of step-by-step laboratory methods and protocols from across the life and biomedical sciences. Each protocol is provided in the Springer Protocol format: readily-reproducible in a step-by-step fashion. Each protocol opens with an introductory overview, a list of the materials and reagents needed to complete the experiment, and is followed by a detailed procedure supported by a helpful notes section offering tips and tricks of the trade as well as troubleshooting advice. With a focus on large comprehensive protocol collections and an international authorship, Springer Protocols Handbooks are a valuable addition to the laboratory.

Experimental Protocols in Biotechnology

Edited by

Neelima Gupta

Chhatrapati Shahu Ji Maharaj University, Kanpur, Uttar Pradesh, India

Varsha Gupta

Department of Biotechnology, Chhatrapati Shahu Ji Maharaj University, Kanpur, Uttar Pradesh, India

Humana Press

Editors
Neelima Gupta
Chhatrapati Shahu Ji Maharaj University
Kanpur, Uttar Pradesh, India

Varsha Gupta
Department of Biotechnology
Chhatrapati Shahu Ji Maharaj University
Kanpur, Uttar Pradesh, India

ISSN 1949-2448 ISSN 1949-2456 (electronic)
Springer Protocols Handbooks
ISBN 978-1-0716-0606-3 ISBN 978-1-0716-0607-0 (eBook)
https://doi.org/10.1007/978-1-0716-0607-0

This Humana imprint is published by the registered company Springer Science+Business Media, LLC part of Springer Nature.
The registered company address is: 1 New York Plaza, New York, NY 10004, U.S.A.

Dedication

Dedicated to all biotechnologists who strived hard during their lives to explore and develop bio-techniques for the cause of science and society

Preface

This publication of protocol manual is an outcome of intense workshop "National workshop on Advanced Molecular Biotechniques" conducted by some of the best minds in India. The approach that some of the researchers took in their experiments, along with their problem-solving style was commendable and was appreciated by the science fraternity.

Looking at the response, the organizers decided to extend the standardized protocols to all the researchers and scientists working in similar fields. This experimental protocol manual covers many important technologies which are routinely used in the laboratories across the world.

We have made an attempt to comprehensively cover different aspects of various techniques, starting from the introduction, the importance of the technique, areas where the technique can be utilized, principle, working, the requirements and standardized methodology. This is different from other protocol series in its essence of showing the possible outcomes after experimentation. It also has notes on troubleshooting that explains possible precautions and guides the researchers in seamlessly executing the experiments.

The protocol manual starts with microbial infection detection in plants with elaboration of the technique. Chapter 1 has standardized protocols for screening and detection of *Ralstonia solanacearum*, (a Gram-negative phytopathogen) a causative agent of bacterial wilt disease in Solanaceae crops leading to severe agroeconomic losses. Chapter 2 elaborates detection of *Fasciola* infection in livestock production. Elaborate protocols for antigen extraction and detection from stool are provided which gives an estimate about pathogen load.

In situ imaging, both in live and fixed zebrafish embryos using cytomegalo virus (CMV) promoter-driven expression of fusion proteins, is discussed in Chapter 3. The chapter presents protocols for plasmid selection and microinjection, screening of embryos having clones expressing the fusion proteins, immunostaining and imaging of zebrafish embryos to study localization and dynamics of proteins of interest encoded by plasmid vectors.

With growing population and restriction of usable agricultural areas, aquaculture and fisheries hold high prospects in future. Chapter 4 elaborates techniques to monitor hematological parameters to assess the health status for diagnosis and prevention of diseases in fish. As the techniques of hematology are different in fishes and mammals, assessment of fish blood parameters is demonstrated with complete description of the techniques and methods employed for the identification of blood parasites and heme indices.

Nematode isolation and detection are challenging. They are responsible for severe infections in animals and plants. In Chapter 5, electron microscopic examination procedures along with molecular taxonomy of roundworms to segregate taxa from arthropod-parasitic, freshwater, marine water, and soil habitats have been elaborated.

Chapter 6 focuses on various computational approaches in the process of drug discovery. It covers various methods and algorithms of computational biology which can be used for structural predictions of lead compound. The chapter not only covers screening and docking methodologies but also elaborates pharmacophore mapping with discussions about metabolism and elimination of drugs.

Antibiotic resistance is posing challenge for scientific and medical fraternity. Chapter 7 deals with the study of microbial biofilms, their diversity as well as genetic complexity which leads to microbial resistance to biocontrol agents such as antimicrobial compounds. It elaborates effective and safe antibacterial alternative strategies as using bacteriophages in an era of rapidly emerging antibiotic resistance.

Chapter 8 elaborates non-radioisotopic safe and effective protocols for studying reversible phosphorylation and dephosphorylation reactions in plant system. The protocols can be extended to any other experimental system. Detailed methodology for in vivo phosphorylation detection is discussed with standardized protocols using western blotting and probing by anti-phosphoantibodies.

Chapter 9 elaborates isolation of plasma membrane and tonoplast-enriched fractions. It provides standardized methodology for estimation of ATPase hydrolytic activity with an important role of possible inhibitory influences of molecules which obstructs its hydrolytic activity.

Chapter 10 explains detection, quantification, and annotation of microRNAs (miRNAs). The conserved miRNA family members play an important function in a number of physiological processes with an essential role in the control of gene expression.

Cell analysis and identification have become easier with advancement in technology. Chapter 11 covers working protocols for isolation, purification, and characterization of natural killer cells (NK cells) from mouse spleen and bone marrow.

Powerful technologies of detection of gene expression which have completely overshadowed northern blotting are discussed in Chapters 12 and 13. Chapter 12 elaborates RNA isolation, quality and quantity evaluation, reverse transcription, and real-time PCR reaction with example of quantification and normalization of gene expression. Chapter 13 elaborates versatile and emerging technology of digital droplet PCR and its applications.

Understanding enzymes, its kinetics, determination of V_{max} and the role of inhibitor have always been challenging for researchers. Chapter 14 brings standardized protocol of alkaline phosphatase activity determination and gradually guides researcher to determine V_{max}, K_M, influence of inhibitors and positive-negative feedback loops.

Lastly, Chapter 15 deals with the introduction and evaluation of non-conserved long non-coding RNA (LncRNA). LncRNA is increasingly finding importance as an important regulator of gene expression. Here detection of lncRNA from tissue section is elaborated.

The protocols compiled are ready-to-use methods which are standardized from existing protocols. We hope that the researchers will find these protocols useful and will be able to tap years of experience of researchers through this book.

Kanpur, India — *Neelima Gupta*
Kanpur, India — *Varsha Gupta*

Acknowledgements

We would like to start by praising and thanking the Almighty for giving us the inspiration, guiding us through the entire journey, and making the experience memorable.

We express our heartfelt and sincere gratitude to everyone who has contributed selflessly to the growth and development of science. Unfailing positive spirit despite the failures and struggles of the scientists all over the world has kept the quest for knowledge alive and laid the foundation stone as well as paved our path into the technological world. Reflections of their knowledge in writings, books, publications, and other documents have helped and inspired us to capture some of the knowledge and pass it on to readers.

We also want to express our gratitude to our parents for their love, blessings, and continuous selfless support, which have made us what we are today, and to our siblings for providing continuous encouragement and support. Thanks are due to Meghana and Agastya who spent their time and helped us with writings and diagrams.

Last but not least, we would like to thank all the colleagues, friends, and students for their support and help during the preparations. We sincerely acknowledge the untiring support of the entire Springer team for their cooperation and support. Special mention must be made to Ms. Aakanksha Tyagi for taking all the trouble to help us whenever required. Her continuous guidance and support made this possible. We would also like to thank Mr. John Martyn and Ms. Padmapriya for all the cooperation and support. Thanks to our spouses for their immense patience and being a constant source of inspiration throughout the journey.

Contents

Editors and Contributors

About the Editors

NEELIMA GUPTA received her M.Sc. (Gold Medalist), M.Phil., and Ph.D. (in 1981) (Zoology) from the AMU, Aligarh, and obtained her D.Sc. degree (Animal Science) from the MJP Rohilkhand University, Bareilly, in 2012. Presently, she works as Vice Chancellor of CSJM University, Kanpur, India, since 2018. She is the recipient of the Ek Janaki Ammal National Award for Animal Taxonomy (highest state award from the Ministry of Environment & Forests), Saraswati Samman and Vigyan Ratna Award (Council of Science & Technology, UP). She has worked extensively on parasite taxonomy, aquatic toxicology of the Ramganga and Ganga Rivers, and fish health. She was nominated by the British Council, DST, and INSA in various academic programs.

VARSHA GUPTA received her Ph.D. (Life Sciences) from Jawaharlal Nehru University, New Delhi, in 2003. She then joined the National Centre for Plant Genome Research, New Delhi, for her postdoctoral training on plant genomics. Presently, she works at Chhatrapati Shahu Ji Maharaj University, Kanpur, since 2006. She has done extensive work on molecular diagnostics using genomic and proteomic approaches and has published several research articles and one textbook on *Basic and Applied Aspects of Biotechnology*.

Contributors

MONA A. AWAD • *Medical Division, Department of Clinical and Chemical Pathology, National Research Centre, Cairo, Egypt*

AASHAQ HUSSAIN BHAT • *Nematology Laboratory, Department of Zoology, Chaudhary Charan Singh University, Meerut, India*

VISHAL CHAND • *Rheumatology Laboratory, Department of Biotechnology, Chhatrapati Shahu Ji Maharaj University, Kanpur, Uttar Pradesh, India*

ASHOK KUMAR CHAUBEY • *Nematology Laboratory, Department of Zoology, Chaudhary Charan Singh University, Meerut, India*

SUBHASH C. CHAUHAN • *Department of Immunology and Microbiology, School of Medicine, University of Texas Rio, McAllen, TX, USA*

ASMITA DAS • *Department of Biotechnology, Delhi Technological University, New Delhi, India*

KYLE A. DOXTATER • *Department of Pharmaceutical Sciences, College of Pharmacy, University of Tennessee Health Science Center, Memphis, TN, USA; Department of Immunology and Microbiology, School of Medicine, University of Texas Rio, McAllen, TX, USA*

ANAMIKA DWIVEDI • *Rheumatology Laboratory, Department of Biotechnology, Chhatrapati Shahu Ji Maharaj University, Kanpur, Uttar Pradesh, India*

ABHISHEK GUPTA • *I3Consulting Pvt. Ltd. Sector 127, Noida, Uttar Pradesh, India*
DILEEP K. GUPTA • *Department of Zoology, Bareilly College, Bareilly, India*
KIRTI GUPTA • *Department of Biological Sciences, Tata Institute of Fundamental Research, Mumbai, India*
NEELIMA GUPTA • *Chhatrapati Shahu ji Maharaj University, Kanpur, India*
VARSHA GUPTA • *Rheumatology Laboratory, Department of Biotechnology, Chhatrapati Shahu Ji Maharaj University, Kanpur, Uttar Pradesh, India*
MEENA JAGGI • *Department of Immunology and Microbiology, School of Medicine, University of Texas Rio, McAllen, TX, USA*
ABHIMANYU KUMAR JHA • *Institute of Technology and Management, Meerut, Uttar Pradesh, India; Department of Biotechnology, Faculty of Life Sciences, Institute of Applied Medicine and Research, Ghaziabad, Uttar Pradesh, India*
SHILPA DESHPANDE KAISTHA • *Department of Microbiology, Institute of Biosciences and Biotechnology, CSJM University, Kanpur, India*
DINESH KUMAR • *Rheumatology Laboratory, Department of Biotechnology, Chhatrapati Shahu Ji Maharaj University, Kanpur, Uttar Pradesh, India*
VINEET KUMAR • *Department of Microbiology, Institute of Biosciences and Biotechnology, CSJM University, Kanpur, India*
SANDEEP K. MALHOTRA • *Parasitology Laboratory, Department of Zoology, University of Allahabad, Allahabad, Uttar Pradesh, India*
ANDREW E. MASSEY • *Department of Pharmaceutical Sciences, College of Pharmacy, University of Tennessee Health Science Center, Memphis, TN, USA; Department of Immunology and Microbiology, School of Medicine, University of Texas Rio, McAllen, TX, USA*
SHAHLA NIGAR • *Centre of Excellence Laboratory, Department of Animal Science, M.J.P. Rohilkhand University, Bareilly, Uttar Pradesh, India*
SIPAHEE LAL PATEL • *Rheumatology Laboratory, Department of Biotechnology, Chhatrapati Shahu Ji Maharaj University, Kanpur, Uttar Pradesh, India*
RAJESH KUMAR PATHAK • *School of Agricultural Biotechnology, Punjab Agricultural University, Ludhiana, Punjab, India*
JAYA PRAKASH • *Community Health Centre, Kanpur, Uttar Pradesh, India*
PAYAL SINGH RAGHUVANSHI • *Rheumatology Laboratory, Department of Biotechnology, Chhatrapati Shahu Ji Maharaj University, Kanpur, Uttar Pradesh, India*
AASHA RANA • *Nematology Laboratory, Department of Zoology, Chaudhary Charan Singh University, Meerut, India*
SADHANA SINGH SAGAR • *Water Quality Division, Center for Water Resource Development and Management, Kozhikode, Kerala, India*
SAID I. SHALABY • *Medical Division, Department of Complimentary Medicine, National Research Centre, Cairo, Egypt*
LATA ISRANI SHUKLA • *Department of Biotechnology, School of Life Sciences, Pondicherry University, Kalapet, Puducherry, India*
DEV BUKHSH SINGH • *Department of Biotechnology, Institute of Biosciences and Biotechnology, Chhatrapati Shahu Ji Maharaj University, Kanpur, India*
MAHENDRA SONAWANE • *Department of Biological Sciences, Tata Institute of Fundamental Research, Mumbai, India*
NEERJA SRIVASTAVA • *Department of Biochemistry, Institute of Biosciences and Biotechnology, Chhatrapati Shahu Ji Maharaj University, Kanpur, Uttar Pradesh, India*

MANISH K. TRIPATHI • *Department of Immunology and Microbiology, School of Medicine, University of Texas Rio, McAllen, TX, USA*

BAISHNAB CHARAN TRIPATHY • *School of Life Sciences, Jawaharlal Nehru University, New Delhi, India*

PRAMILA DEVI UMRAO • *Department of Microbiology, Institute of Biosciences and Biotechnology, CSJM University, Kanpur, India*

ANSHIKA YADAV • *Parasitology Laboratory, Department of Zoology, University of Allahabad, Allahabad, Uttar Pradesh, India; Wild Life Section, Zoological Survey of India, Kolkata, West Bengal, India*

MURALI M. YALLAPU • *Department of Immunology and Microbiology, School of Medicine, University of Texas Rio, McAllen, TX, USA*

Chapter 1

Enzyme-Linked Immunosorbent Assay Detection of Bacterial Wilt–Causing *Ralstonia solanacearum*

Pramila Devi Umrao, Vineet Kumar, and Shilpa Deshpande Kaistha

Abstract

Ralstonia solanacearum, a Gram-negative phytopathogen, is the causative agent of a devastating bacterial wilt disease in Solanaceae crops leading to severe agroeconomic losses. Detection of bacterial wilt causing *Ralstonia solanacearum* from infected plants and water or soil samples is critically required for the early diagnosis, prophylaxis, and treatment of the disease. Enzyme-linked immunosorbent assay (ELISA) is one of the most proficient and relevant methods to detect the phytopathogen because of the specific interaction of enzyme linked antibodies with pathobiogenic molecules. Immunosorbent assay is a microtiter plate--based protocol designed for detecting and quantifying infected plants/soil samples select pathogen or its pathogenicity factor. Serodiagnostic techniques are highly beneficial over traditional methods as they are fast, reliable, easy, and inexpensive. Methodologies to prepare sample culture, plant materials, and soil for the detection of bacterial wilt causing *Ralstonia solanacearum*; standardized ELISA protocols; and specific precautions to carry out the assay successfully are discussed in this chapter. The sensitivity of ELISA test for different samples is discussed in the validation and interpretation section.

Keywords *Ralstonia solanacearum*, ELISA, Antibodies, Race, Biovar, Bacterial wilt detection

1 Introduction

Ralstonia solanacearum is a Gram-negative, aerobic, motile, non-spore forming and noncapsulating bacterial phytopathogen [1, 2]. Bacterial wilt caused by *R. solanacearum* is one of the most problematic diseases of many economically important crops belonging mainly to Solanaceae family such as potato, tomato, pepper, and eggplants [1, 3–5], and nonsolanaceous crops such as ginger, banana, and groundnut [3, 6], including ornamental plants distributed worldwide. Due to heterogeneity, species complexity and disease severity the pathogen has been ranked second after *Pseudomonas syringae* in top ten pathogens categorized [7]. Based on geographical location and biomolecular typing, four phyllotypes are described for *R. Solanacearum*, namely, phyllotype I strains in Asia, phyllotype II strains in the Americas; phyllotype III strains in

Neelima Gupta and Varsha Gupta (eds.), *Experimental Protocols in Biotechnology*, Springer Protocols Handbooks, https://doi.org/10.1007/978-1-0716-0607-0_1,

Africa; and phyllotype IV strains in Indonesia [8]. This phytopathogen typically infects through roots using natural openings or wounds inflicted by mechanical devices and insect or nematode bites [9]. The pathogen then moves up the stem and colonizes the vascular bundles in the xylem tissue eventually causing irreversible wilting in the host plant. The pathogen may be isolated from the whitish ooze from infected stem cutting or from infected soil. Its control is challenging as the pathogen grows endophytically, transmitted by water, and survives in soil for long durations of time [10].

Control of agricultural losses due to phytopathogens constitutes one of the major thrust areas of research of which diagnostics is an integral part to ensure early detection and therapeutic measures to be implemented [11]. Hence, different laboratories in the world are working toward developing an easy and rapid diagnostic technique for *R. solanacearum* which may be useful in the laboratory as well as on the field [10, 12–16]. ELISA-based detection assays are most reliable due to high specificity of antigen–antibody interaction, rapid execution, ease of performance as well as cost-effectiveness to understand bacterial wilt disease epidemiology and quantification [17, 18].

2 Techniques

Serodiagnosis is an effective technique for rapid, economical, and easy monitoring of soil and seeds for good agricultural practices. Enzyme linked immunosorbent assay (ELISA) was first described as a solid immunoassay to detect the presence of antigen using antibody-coated surfaces [19]. The basic principle of ELISA is that antigen–antibody interactions are performed on an adsorbent surface such as nitrocellulose membrane (NCM ELISA), polystyrene microtiter plates, or polyvinylpyrrolidone plates for detection and quantification of the antigen in a heterogeneous mixture. Enzyme-conjugated antibodies react with antigen–antibody complex, and any excess is removed by a washing step. An enzyme-specific substrate is then added to the wells. The bound enzyme-conjugated antibody plays a role in hydrolyzing a colorless substrate to give colored, fluorescent or chemiluminescent end products which can be measured visually (qualitative test) or by using spectrophotometry (quantitative test). Development of color intensity is dependent on number of catalyst enzymes bound to the antigen–antibody complex. Variant ELISA include indirect ELISA, competitive ELISA, and double antibody sandwich ELISA (DAS ELISA) [20].

Indirect ELISA is a two-step binding modification of direct ELISA to improve sensitivity of the assay. Indirect ELISA assay uses direct adsorption of antigen/sample onto the immunosorbent surface as the first step in the assay. The next step involves binding of surface-bound antigen by its specific primary antibodies. Unbound primary antibodies are washed and detection of antigen-bound primary antibodies is carried out with multiple enzyme-linked secondary antibodies attached to the constant region of the primary antibody. Unbound secondary antibodies are again washed before chromogenic/detector substrates are added for indirect detection of antibodies bound to the antigen on the substrate. The advantages of indirect ELISA are the use of a polyclonal enzyme-labeled secondary antibody that increases signal intensity, flexibility in use, and high economical viability in the laboratory when multiple ELISAs are to be performed. Disadvantages include cross-reactivity and nonspecific positive reaction with the secondary antibody.

Competitive or inhibition ELISA for antigen detection uses the principle of competitive binding between the test antigen and labeled antigens of known concentrations to the coated and immobilized antibody. Make a mix of biotin-labeled antigen and unlabeled antigen (test and standard reference). This mixture is now allowed to incubate competitively with previously coated antibody on the immunosorbent surface. Postincubation, unbound antigen is washed off with phosphate buffered saline tween (PBST). Secondary antibodies with horseradish peroxidase conjugated to streptavidin are used for the binding to biotin-labeled antigen and development of color. Biotin specifically binds to streptavidin, which is conjugated to the enzyme and increases the sensitivity of detection. The more the concentration of test antigen, the lower the color intensity in the ELISA well. This assay is typically used for measurement of low molecular weight antigenic haptens such as peptides and steroids.

In DAS ELISA, a capture antibody coated on the microtiter well binds to a test antigen in a sample. Enzyme-conjugated pathogen-specific antibody is then added to this complex, creating a sandwich of the test antigen between the two specific antibodies. Primary monoclonal antibodies specific to the target antigen(s) are derived from one species, and enzyme-conjugated secondary antibodies show specific binding against primary antibodies (IgG's Fc) which are prepared in another animal species. Advantages of DAS ELISA include high sensitivity as two pairs of coating and primary antibody are used for capture and detection. This technique also allows for versatility of sample detection without cross-reactive signals. This chapter will focus on DAS ELISA as the immunodiagnostic technique for the detection of the phytopathogen.

Secondary antibodies bound to monoclonal primary antibody are usually used for the measurement of antigen by conjugating them with an enzyme system, fluorescent dyes, or luminescent biomolecules. The use of enzyme-linked antibodies is popular in most laboratories as it requires only a visible spectrophotometer

Table 1
Commonly used enzyme systems in chromogenic ELISA

Enzyme system	Substrate	Product	Detection limit of substrate	Advantages	Disadvantages
Alkaline phosphatase (AP)	pNPP (*p*-nitrophenyl phosphate)	*p*-nitrophenol (yellow) at 460 nm	100 ng/ml 1°Ab—1:500 2°Ab—1:5 K,1:20 K	Powder and ready-to-use solutions available Nonhazardous	Lower sensitivity compared to alternative substrates
Horseradish peroxidase (HRP)	TMB (3,3,5,5′ tetramethylbenzidine) and H_2O_2	Blue-colored product which may turn yellow on adding stopping agents such as dil phosphoric or sulfuric acid	20 pg/ml to 80 pg/ml	Most sensitive ELISA substrate Higher turnover rate than AP resulting in fast Color development Very versatile Available in several forms: Powder, tablet, ready to use	Carcinogenic Require spectrophotometry Requires stopping agent (dil sulfuric acid)
Horseradish peroxidase (HRP)	ABTS (2,2-Azino bis [3-ethylbenzothaizoline-6-sulfonic acid]-diammonium salt	Green color at 420 nm	2.5 ng/ml 1°Ab—1:1 K 2°Ab—1:5 K to 1:50 K	Ready-to-use substrate Also used to detect antioxidant activities	Not as sensitive as TMB or OPD
Horseradish peroxidase (HRP)	OPD (*O*-phenylenediamine dihydrochloride)	Yellow orange color	70 pg/ml 1°Ab—1:1 K 2°Ab—1:5 K to 1:50 K	Lower sensitivity than TMB. Higher than ABTS	Not available in ready to use format
β-Galactosidase	ONPG (ortho-nitrophenyl-β-galactoside)	Water-soluble yellow product	100 ng/ml 1°Ab—1:500 2°Ab—1:5 K	Available in powder form	Protect from moisture and light

1°Ab: Primary antibody, 2°Ab: Secondary antibody
Adapted from information at www.info.gbiosciences.com

and inexpensive reagents that are easily available. Table 1 summarizes the different enzyme systems that are readily used for protocols aimed at diagnoses of diseases/phytopathogens.

Phytopathogen detection ELISA requires only skill in execution, with no special microbiology training nor the presence of sophisticated molecular instrumentation such as polymerase chain reaction (PCR) machine. The disadvantage of the method is that its does not detect viability of the phytopathogen but only the presence of biomolecules that are antibody specific and sensitivity is lower in comparison to molecular methods such as PCR. Here, we describe methods for sample preparation from plant, soil and culture; test procedure; and detection methods of DAS ELISA.

3 Materials

All media are procured from HiMedia, India, and all analytical grade chemicals are procured from Merck, India.

1. Standard reference strain—Ralstonia solanacearum F1C1 for positive control.
2. Culture media 2,3,5-triphenyl tetrazolium chloride (TZC) for *R. solanacearum* [21].

Constituents	Concentration
CAS amino acids/casein hydrolysate	01 g/l
Peptone	10 g/l
Glucose	05 g/l
Agar (for solid media only)	17 g/l
1% TZC solution (after autoclaving)[a]	05 ml

[a]1% stock solution of 2,3,5-triphenyl tetrazolium chloride can be filter-sterilized or autoclaved for 5 min at 121 °C, and stored at 4 °C or frozen)

3. Modified Selective Media 1 (SM1): Semiselective media for enrichment of *R. solanacearum* from soil and asymptomatic plants [22].

Constituents	Concentration
Casamino acids/casein hydrolysate	01 g/l
Peptone	10 g/l
Glucose	05 g/l
Agar (for solid media only)	17 g/l
1% TZC solution (after autoclaving)[a]	05 ml

(continued)

Constituents	Concentration
Antibiotic solution[b]	
Polymyxin B	100 mg/l
Bacitracin	25 mg/l
Chloramphenicol	05 mg/l
Cycloheximide (antifungal)	100 mg/l
Crystal violet	05 mg/l

[a]1% stock solution of 2,3,5-triphenyl tetrazolium chloride can be filter-sterilized or autoclaved for 5 min at 121 °C, and stored at 4 °C or frozen
[b]Prepare stock solution in ethanol and store in foil-covered bottle at 4 °C. Alternatively prepare fresh in 70% ethanol 30 min prior to use

4. Modified SMSA medium [23].

Constituents	Concentration
CAS amino acid/casein hydrolysate)	01 g/l
Peptone	10 g/l
Glycerol	05 ml
Agar (for solid media only)	17 g/l
Crystal violet	05 mg/l
Polymyxin ß sulfate	100 mg/l
Bacitracin	25 mg/l
Chloromycetin	05 mg/l
Penicillin G	0.5 mg/l
Cycloheximide (antifungal)	100 mg/l

Adjust pH to 6.5–7.0 if necessary
Autoclave at 121 °C for 20 min. After autoclaving cool the medium to 55 °C and add 5 ml of a 1% stock solution of 2,3,5-triphenyl tetrazolium chloride. The stock can be filter-sterilized or autoclaved for 5 min at 121 °C, and stored at 4 °C or frozen
Dissolve in 5 ml of 70% ethanol 30 min prior to use
On solid medium, colonies of *R. solanacearum* usually are visible after 2–5 days of incubation at 28 °C. Typical bacterial colonies appear fluidal, irregular in shape, and white with pink centers

5. Developed single chain variant antibody (primary antibodies) and secondary antibody horseradish peroxidase enzyme conjugates (ELISA Kit Cat. No. SRP 33900 [Agdia, www.agdia.com] purchased from Life Technologies Pvt. Ltd., India., available commercial kits).
6. Developing reagents: TMB (3,3′,5,5′ tetramethyl benzidine), hydrogen peroxide (H_2O_2).

7. Buffer Formulations (1×)

(a) Extraction Buffer.

Constituents	Concentration
Sodium sulfite (anhydrous)	1.5 g
Polyvinylpyrrolidone (PVP) MW 40,000	20.00 g
Sodium azide	0.2 g
Powdered egg (chicken) albumin, grade II	2.0 g
Tween 20	20.0 g
DDW Adjust pH to 7.4	1.0 l

(b) Carbonate Coating Buffer.

Constituents	Concentration
Sodium carbonate (anhydrous)	1.59 g
Sodium bicarbonate	2.93 g
Sodium azide	0.2 g
DDW Adjust pH to 9.6. Store at 4 °C	1.0 l

(c) PBST Buffer (Wash Buffer).

Constituents	Concentration
Sodium chloride	8.0 g
Sodium phosphate, dibasic (anhydrous)	1.15 g
Potassium phosphate, monobasic (anhydrous)	0.2 g
Potassium chloride	0.2 g
Tween 20	0.5 g
DDW	1 l

Adjust pH to 7.4

The components of all buffers are dissolved in 1-l double distilled water (DDW). It is advisable to use freshly prepared buffer. Store developing regents at 4 °C and bring reagents to normal temperature before use

(d) Stopping Solution.

Sulfuric acid	0.2 M

8. Equipment.

(a) Incubator at 37 °C.
(b) Microtiter plate reader.
(c) Centrifuge.
(d) Micropipettes and sterile tips.

4 Methods

4.1 Preparation of Sample from Plant Material, Bacterial Culture, and Infected Soil

4.1.1 Plant Materials

Sample collection of the plant sample depends upon infection strategy and type of pathogen as well as plant and tissue type. *Ralstonia solanacearum* is a vascular pathogen; hence, it is best to choose crown-region node of symptomatic plants. The pathogen can be isolated from root, stem, tuber, petioles, and leaf tissue of asymptomatic and symptomatic plants.

Nonsymptomatic Plants (Fig. 1a)

1. Enrich extracts by inoculating in semiselective SMSA broth for 48 h at 30 °C [16]. Enriched stem extracts can be further processed as samples in extraction buffer for DAS-ELISA. In order to increase sensitivity, neutralization of tissue proteins may be carried out using specific tissue extraction buffers.
2. Proceed after enrichment as described for symptomatic plant samples.

Symptomatic Plants (Fig. 1b)

1. Collect bacterial ooze sample from symptomatic plants showing bacterial wilt symptoms by cutting 10 cm section of plant material.
2. Wash and sterilize with 0.5% NaOCl for 3 min followed by 70% ethanol.
3. After proper washing and drying, prepare 5 mm section pieces, suspend in extraction buffer, and homogenize the tissue by grinding.
4. Centrifuge sample at 358 × g for 3 min to spin down plant debris.
5. Aspirate the supernatant into fresh sterile tubes.

Potato Tuber/Slice (Fig. 1c)

1. For potato tuber/slice testing, prepare a 5 mm deep cross section from infected tuber and scoop the tissue into a petri dish.
2. Mash tuber tissue using sterile scalpel and add 1 mL extraction buffer. Reduce extraction buffer based on tissue size to make the protocol sensitive.

3. Vortex to release bacteria into the extraction buffer.
4. Centrifuge sample at 358 × g for 3 min to spin down tuber debris.
5. Aspirate the supernatant into fresh tubes.
6. Prepare serial dilution of the sample using buffer for use in DAS-ELISA. Mix well before and after serial dilution by drawing up and down using pipette or shaking or tapping bottom of tube or rolling between palms while wearing a ring.
7. Preparation of tenfold serial dilutions: transfer 0.1 ml vortexed culture to 0.9 ml saline (10^{-1} dilution), vortex, then transfer 0.1 ml of 10^{-1} dilution to 0.9 ml saline (10^{-2} dilution), vortex—repeat this process until you have achieved 10^{-8} dilution.

Soil-

1. Prepare soil suspension by dissolving 10 g soil in 100 ml 1× phosphate buffer saline agitated at 140 rpm for 30 min and wait for settling down of soil and particulate matter.
2. Perform tenfold serial dilution of soil suspensions using buffer to improve the detection efficiency in DAS-ELISA as described previously.
3. The supernatant may be selectively enriched by SMSA broth to enhance sensitivity as described previously if count is too low for detection.

Culture (provided by Dr. Suvendra Kumar Roy, Head, Department of Molecular Biology and Biotechnology, Tezpur University, Assam, India) (Fig. 1d)

1. For identification of bacteria, grow isolated bacterial strains on tryptone soy broth (TSB) or TZC media as described previously.
2. Prepare tenfold serial dilution of culture in sterile saline to obtain approximately 10^5 to 10^8 cfu/ml (measuring by optical density at 620 nm using spectrophotometer).

4.2 DAS-ELISA Protocol

4.2.1 Principle

Double antibody sandwich protocol known as DAS-ELISA is widely used for identification of *Ralstonia solanacearum* [10, 15, 17, 18]. Captured antibodies are coated to the test wells of a polystyrene microtiter plate followed by blocking with 3% BSA. Test samples are then added to the antibody-coated microtiter plate wells. *R. solanacearum* pathogen or its exopolymeric substance in the test sample is captured by coating antibodies on the microtiter plate during the incubation period. After incubation, the plate is washed to remove unbound sample. An enzyme peroxidase conjugate solution of antibody is added and bound to any captured *R. solanacearum*. After incubation, the plate is washed to remove

Fig. 1 Samples for detection of bacterial wilt caused by *Ralstonia solanacearum*. (**a**) Asymptomatic plants. (**b**) Symptomatic plants. (**c**) Infected tuber/potato slice (inset). (**d**) *Ralstonia solanacearum* F1C1 pure culture on 2,3,5-triphenyl tetrazolium chloride (TZC) media

any unbound conjugates. Subsequently, TMB substrate is added to the microtiter plate so that the substrate is converted by the enzyme to produce a colored product. TMB is a colorless and reduced form of substrate. In the presence of hydrogen peroxide and peroxidase enzyme, two hydrogen atoms from H_2O_2 are used to oxidize TMB turning it blue. Hence, the blue color produced indicates the presence of EPS for the production of peroxidase conjugates. High concentration of peroxidase causes blue coloration to precipitate and hence stopping solution is used. When sulfuric acid is used as stopping solution, a yellow colored solution is obtained. The color change from biochemical reactions can be observed visually or measured with a spectrophotometer scanning at different wavelengths. The major advantage of DAS-ELISA is that it is suitable for crude antigen preparations obtained from soil or plant tissue without requiring any purification steps or affecting specificity and sensitivity.

4.2.2 Protocol (Flowchart Is Presented in Fig. 2)

1. Prepare a layout of the microtiter plate both on paper as well on the software you may be using (Fig. 3). Many multititer plate readers and ELISA software allow this to be done online. Also, using a waterproof marker, designate clearly wells for standard curve, different blanks, positive control, and test samples with their dilution. Making a practice of consistently placing controls/test samples on the same wells will ensure that no manual errors creep in when actually setting up the experiment.

Coat captured antibodies on test wells of a polystyrene microtiter plate

↓

Incubate coated plates inside the humid box overnight at 4 C

↓

Block test wells with 3% BSA in 1XPBS for 30'xRT

↓

Discard blocking buffer by flicking microtiter plate

↓

Dispense samples (test and reference) into wells and make dilutions

↓

Incubate plate inside the humid box for 60'xRT or 30'x37°C

↓

Wash plate with 1X PBST about 3 times to remove all unbound Ag

↓

Dispense peroxidase enzyme conjugate captured Ab per testwell

↓

Incubate at room temperature (RT) for 45 min

↓

Wash the plate 3 times with wash buffer (1X PBST)

↓

Add TMB substrate into each test well, wait for 30 min. Add stopping buffer if required

↓

Evaluate result by color change and measurements using spectrophotometer (A_{600})

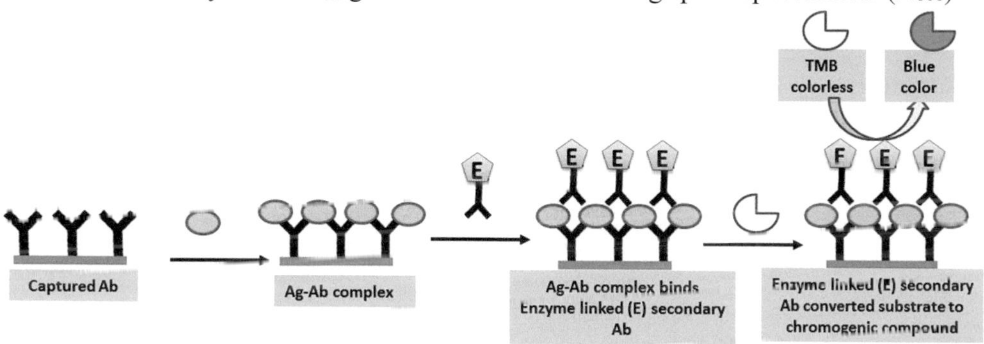

Fig. 2 Schematic flowchart for DAS ELISA protocol

2. Coat each test well of a polystyrene microtiter plate with 100 μl of captured anti-*R. solanacearum* EPS monoclonal antibodies (1:100 in coating buffer). Fill the edge wells of the microtiter plates with sterile distilled water to prevent evaporation and edge effects.
3. Incubate the coated plates inside a humid box 4 °C overnight or 37 °C for 30 min or room temperature (RT) for 60 min.

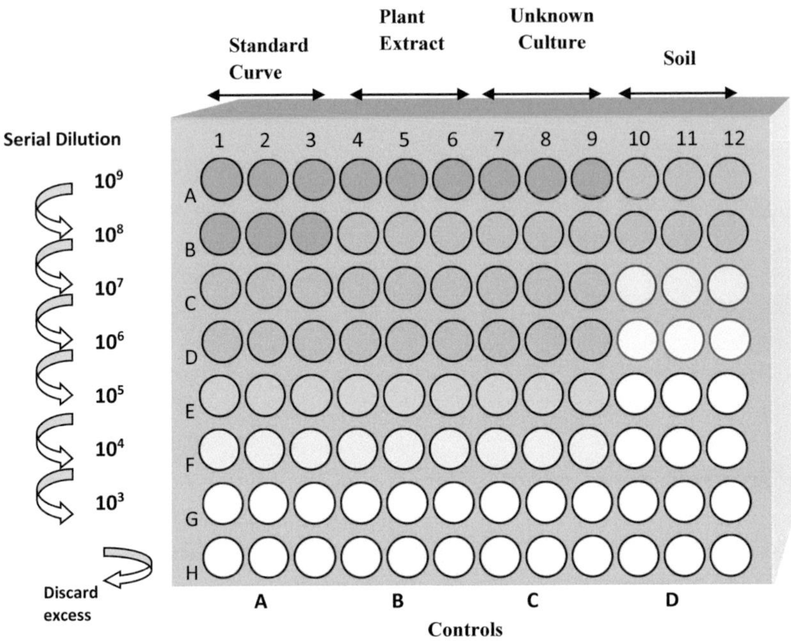

Fig. 3 Schematic design of ELISA microtiter plate assay

Design the humid box by using autoclaved 1 ml tip boxes that are coated with filter paper soaked in sterile water.

4. After coating incubation, remove coating antibodies and block the test wells with 3% BSA in 1× PBS for 30 min to prevent nonspecific binding.
5. Discard blocking buffer by flicking microtiter plates over sink or collecting basin.
6. Add all samples in triplicate. Positive control *R. solanacearum* is set at 10^9 cells/ml to be added to the top most well. For quantification of antigen, tenfold serial dilution is performed in column (1–6 wells). Dispense 110 μl of positive control (*R. solanacearum* F1C1) into positive control wells, and dispense 100 μl broth only for negative control. Add 90 μl buffer to all wells in which dilution is to be made. Transfer using multichannel micropipette 10 μl from the top most well to the next well in the column (1:10). Mix well by pipetting up and down. Again, transfer 10 μl of mix to the next well in the column and mix well. Discard the last 10 μl dilution such that all wells should now contain only 100 μl of sample dilution. Similarly, serial dilutions are made for plant samples, soil, and unknown culture in parallel (Fig. 3).
7. Incubate plates inside the humid box at room temperature (RT) for 60 min or 37 °C for 30 min.

8. After incubation, wash plates with 1× PBST using drop bottle at least three times to remove all unbound antigenic sample. Discard washing buffer by flicking microtiter plate over sink or collecting basin. ELISA plates may also be inverted on blotting paper or newspaper to blot any excess fluid in the wells.
9. Dispense 100 μl of 1× peroxidase enzyme conjugate capture antibody (as per kit guidelines) per test well. Standardize highest dilution of conjugated antibody that gives results with positive control with the reagents to reduce background noise.
10. After 45 min incubation at room temperature, wash the plate three times with wash buffer as described previously.
11. Dispense 100 μl of TMB substrate into each test well and wait for 30 min for color development. Use color blocking agent to stop enzymatic reaction if required.
12. Evaluate the results by examining the test wells for color change and measure by spectrophotometer at 600 nm for blue color or 450 nm for yellow color if using the stopping solution to terminate the reaction.

4.2.3 Validation and Interpretation of Results

Qualitative Analysis

Observe the appearance of blue color by visualization after about 20 min of incubation. Negative wells and blanks should remain virtually clear. Qualitative data set can also be used as a positive/negative test for *R. solanacearum* by comparing with blue coloration observed in positive control. The dilution at which color development has occurred gives an estimate of microbial load. This qualitative data can be used for diagnosis for the presence of phytopathogen even if spectrophotometer is not available.

Quantitative Analysis

For data quantification, use microtiter plate reader for measuring the optical density of all the wells at 600 nm wavelength. The wells in which a blue color develops indicate positive results. Use 450 nm wavelength for measurement of yellow coloration observed upon addition of sulfuric acid as stopping solution. Wells in which there is no significant color development indicate negative result. Test results would be valid only if positive control wells give a positive result and buffer or negative control wells remain colorless.

4.2.4 Assay Sensitivity for Different phytopathogen Sources

The sensitivity of ELISA depends on methodologies used for sample collection strategy in the field, as well as laboratory techniques [16, 17, 24]. ELISA sensitivity for *R. solanacearum* detection is found to be 10^5 to 10^6 CFU/ml, making it relatively insensitive for early detection in asymptomatic plants infected with the pathogen [12]. In soil samples, *R. solanacearum* is typically found in the range of 10^4 to 10^5 CFU per gram soil, which can be increased by applying an enrichment procedure, for example by incubating extracts in a selective broth [11, 18]. In order to increase the sensitivity of detection, selective enrichment using semiselective broth such as SM1 media or Modified SMSA can lead to enhancement in the detection of biovars 1 and 2A in soil using DAS-ELISA by 10–100-fold [10]. To avoid the risk of cross-reactivity with other bacterial populations present in the soil suspensions, the use of monoclonal antibodies, which are commercially available, rather than polyclonal antibodies is prescribed. An ELISA protocol for bacterial wilt pathogen detection from potato sample, wherein tubers can be processed for detection by ELISA has been developed [25]. Bacterial population can also be increased by 10–100 fold by bioindicator host plants of Zingiberaceae and Costaceae family [26]. DAS-ELISA kits, including the sterilized, concentrated (10×) enrichment broth, are available with commercial companies. Recent monoclonal antibodies specific for pathogenicity factors also permit the differentiation of virulent and avirulent strains as well as different biovars. Hence, commercially available ELISA immunodiagnostic kits are useful for the rapid identification of pathogens from different plant materials, soil samples, and cultures. This test can be also used to identify race/biovar such as assigning the pathogen to race 1 or biovar 3 and race 3 biovar 2 [17]. ELISA test is highly useful for rapid and sensitive detection for bacterial wilt in less equipped laboratories concerned with seed health determination and plant quarantine [18, 27].

4.2.5 Generating Standard ELISA Curve and Fitting (Fig. 4)

1. Prepare a standard curve using absorbance data for different cell concentrations of test *R. solanacearum* F1C1 by plotting mean absorbance (*y* axis) against known quantity of antigen/bacterial sample (*x* axis). For quantification assays, a standard curve is crucial for interassay data accuracy and should be run for every plate in order to avoid variations caused by pipetting error, incubations, temperature, and reagent concentrations. Blank values are subtracted from the test values.
2. Draw a best fit curve through the points in the graph.
3. Extrapolate the spectrophotometric absorbance readings of unknown in test samples from the standard curve by extending a horizontal line from *y*-axis to the best-fit curve. At the intersection, trace using a vertical line to the value on the *x*-axis (Fig. 4).
4. To obtain accurate values, multiply the cell number or concentration of bacterial antigen with the dilution factor.
5. Perform validation of ELISA assay by using the checkerboard analysis with large sample size (4–5 per sample) for determining the following parameters [20]:
   (a) *Limit of Detection (LoD)*: Lowest concentration of antigen that can be detected by the assay.

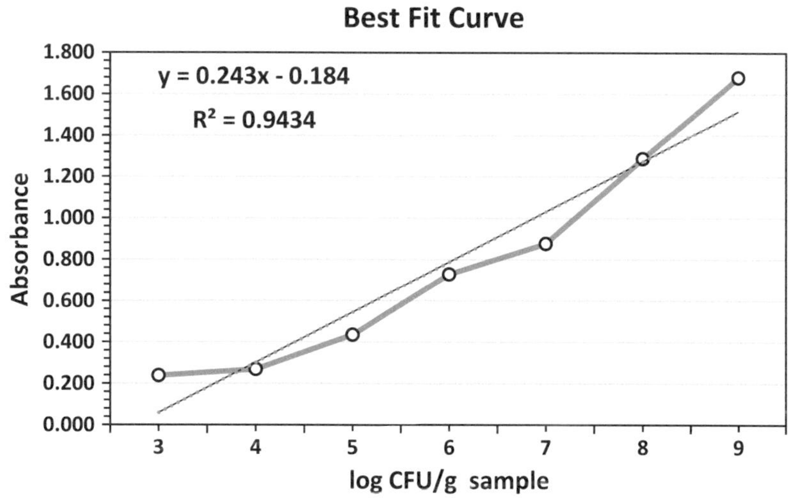

	Absorbance in triplicate				
cell no.	1	2	3	Mean	SD
3 log 10	0.243	0.22	0.251	0.238	0.016
4 log 10	0.255	0.265	0.287	0.269	0.016
5 log 10	0.487	0.398	0.421	0.435	0.046
6 log 10	0.732	0.723	0.731	0.729	0.005
7 log 10	0.899	0.856	0.876	0.877	0.022
8 log 10	1.23	1.29	1.34	1.287	0.055
9 log 10	1.67	1.59	1.78	1.680	0.095
Unknown	0.567	0.523	0.566	0.552	0.025

Fig. 4 Standard curve and plotting best-fit curve for ELISA

(b) *Limit of Quantification (LoC)*: Highest and lowest concentrations of antigen that can be measured with acceptable levels of precision and accuracy.

(c) *Spike Recovery*: Used to identify if diluent has any effect on the measurement. If a known quantity of antigen is added to diluent solution, then it should be possible to get the same amount during measurement. If not, then the diluent buffer may have reacted with the antigen.

(d) *Accuracy*: Closeness of mean test results to reference measurements.

(e) *Precision (Intra assay)*: Closeness in degree of scatter between multiple measurements of the same sample.

Calculate standard deviation and coefficient of variation in order to develop confidence in the generated data.

Intra-assay precision may be calculated using standard deviation (SD) using Excel or GraphPad Prism software or software given with multititer plate readers.

Coefficient of variance (CV) or relative standard deviation (RSD) is used for calculating precision of results between different assays. Usually a CV of 15% is allowed for ELISA.

$$CV = [\text{Standard Deviation}/\text{Mean}] \times 100$$

Checkerboard analysis for amount of antigen, primary antibody, and secondary antibody used for ELISA standardization and optimization is highly recommended as it helps to remove background noise, allows for economical use of antibodies (even as higher concentrations may be recommended as a generalization in ELISA kits), and reduce false positive results.

5 Notes

1. If a quick assay is required where sensitivity is not an issue, the protocol is standardized for 4 h. In order to troubleshoot for sensitivity issues, protocols may be standardized by (a) increasing antibody concentrations, (b) increasing time of incubation, (c) temperature of incubation, and (d) using varying concentrations of antigen (sample). It is important to ensure that antibodies are stored at or below −20 °C and not frequently defrosted [17–19].
2. Place the plates inside a humid box prepared with wet cotton bed to avoid desiccation during incubation steps.
3. Capture antibodies can be re-collected and reused for at least 2 more experiments. Free antibodies are to be collected in glass screw-capped tube and stored at −20 °C. An overnight incubation gives best results in comparison to room temperature of 37 °C incubation. Use 3% BSA as blocking agent to avoid false positive result [19].
4. Assay sensitivity using the same set of secondary antibodies can be improved by increasing incubation time and temperature (up to 37 °C) and the use of a plate shaker.
5. The sensitivity of *Ralstonia solanacearum* depends up on the concentration of cells present in the samples as well as the sample type used for testing [18]. The mentioned *Ralstonia* ELISA test kit detects *R. solanacearum* to the species level and cannot differentiate race or biovar. This test also cannot differentiate pathogenic and nonpathogenic strains. However, some commercial kits also offer biovar specific antibodies and may be purchased based on the specific purpose of performing the ELISA.

6. The lowest sensitivity limit of the described DAS-ELISA is 10^4 to 10^5 cfu/ml *R. solanacearum*. This assay is most sensitive for detecting the pathogen in plant samples as well as culture but not useful for soil samples. *R. solanacearum* detection in the soil needs specific buffer formulations such as citrate buffer followed by enrichment procedure as described previously [11, 15].
7. Excess antigen may also cause the "hook effect" which occurs when antibody is insufficient to bind antigen, resulting in a weak signal. Serial dilution of sample is therefore important in assessing antigen concentrations [25].
8. Air bubbles, present at the time of reading may alter results and hence must be eliminated prior to taking readings. Similarly, ensure that the base of the microtiter plate is clean before placing on the microtiter plate reader.
9. It is essential to prepare all blanks, standard solutions, and test sample in triplicate so that an accurate standard curve can be constructed and statistical analysis can be performed on test data set [20].
10. Interference and potential problems may occur due to handling inaccuracy and cross-contamination of samples during analysis. All the plasticware and glassware must be clean. Inaccurate micropipetting during serial dilution can cause magnification of errors [20].
11. High background readings are due to (a) nonspecific binding, which can be reduced by increasing blocking time or changing blocking reagent. Blocking agents may also be added to the antibody buffers to reduce nonspecific binding. (b) Reduce primary antibody concentration by a checkerboard analysis to determine optimal concentration [20].

References

1. Álvarez B, Biosca EG, López MM (2010) On the life of Ralstonia solanacearum, a destructive bacterial plant pathogen. Technol Educ Top Appl Microbiol Microb Biotechnol 1:267–279
2. Peeters N, Guidot A, Vailleau F, Valls M (2013) Ralstonia solanacearum, a widespread bacterial plant pathogen in the post-genomic era. Mol Plant Pathol 14:651–662. https://doi.org/10.1111/mpp.12038
3. Elphinstone JG, Allen C, Prior P, Hayward AC (2005) The current bacterial wilt situation: a global overview. Am Phytopathol Soc (APS Press St Paul):9–28. https://doi.org/10.1094/9780890544914.009
4. Chandrashekara K, Prasannakumar M (2012) Aggressiveness of Ralstonia solanacearum isolates on tomato. J Exp Sci 3:5–9
5. Ivey MLL, Gardener BBM, Opina N, Miller SA (2007) Diversity of Ralstonia solanacearum infecting eggplant in the Philippines. Phytopathology 97:1467–1475. https://doi.org/10.1094/PHYTO-97-11-1467
6. Kumar A, Sarma YR (2004) Characterization of Ralstonia solanacearum causing bacterial wilt in ginger. Indian Phytopath 57:12–17
7. Mansfield J, Genin S, Magori S et al (2012) Top 10 plant pathogenic bacteria in molecular plant pathology. Mol Plant Pathol 13:614–629. https://doi.org/10.1111/J.1364-3703.2012.00804

8. Fegan MPP (2005) How complex is the Ralstonia solanacearum species complex? In: Allen C, HAC PP (eds) Bacterial wilt: the disease and the Ralstonia solanacearum species complex. American phytopathology society, St. Paul, pp 449–461
9. Hayward A (2000) Ralstonia solanacearum. In: Lederberg J (ed) Encycopedia of microbiology Vol 4. Academic Pres Inc, San Diego, CA, pp 32–42
10. Shahbaz MU, Mukhtar T, Ul-Haque IM, Begum N (2015) Biochemical and serological characterization of Ralstonia solanacearum associated with chilli seeds from Pakistan. Int J Agric Biol 1560–8530
11. Pradhanang PM, Elphinstone JG, Fox RTV (2000) Sensitive detection of Ralstonia solanacearum in soil: a comparison of different detection techniques. Plant Pathol 49:414–422. https://doi.org/10.1046/j.1365-3059.2000.00481.x
12. Fang Y, Ramasamy RP (2015) Current and prospective methods for plant disease detection. Biosensors 5:537–561. https://doi.org/10.3390/bios5030537
13. Döölotkeldieva T, Bobuşeva S (2014) Identification and prevalence of Ralstonia solanacearum from potato fields of Kyrgyzstan. Manas J Agric Vet Life Sci 4:1
14. Sikirou R, Beed F, Ezin V et al (2017) Distribution, pathological and biochemical characterization of Ralstonia solanacearum in Benin. Ann Agric Sci 62:83–88. https://doi.org/10.1016/j.aoas.2017.05.003
15. Priou S, Gutarra L, Aley P (2006) An improved enrichment broth for the sensitive detection of Ralstonia solanacearum (biovars 1 and 2A) in soil using DAS-ELISA. Plant Pathol 55:36–45. https://doi.org/10.1111/j.1365-3059.2005.01293.x
16. Priou S, Gutarra L, Aley P et al (2010) Detection of Ralstonia solanacearum (biovar 2A) in stems of symptomless plants before harvest of the potato crop using post-enrichment DAS-ELISA. Plant Pathol 59:59–67. https://doi.org/10.1111/j.1365-3059.2009.02155.x
17. Behiry SI, Mohamed AA, Younes HA et al (2018) Antigenic and pathogenicity activities of Ralstonia solanacearum race 3 biovar 2 molecularly identified and detected by indirect ELISA using polyclonal antibodies generated in rabbits. Microb Pathog 115:216–221. https://doi.org/10.1016/J.micpath.2017.12.060
18. Rajeshwari N, Shylaja M, Krishnappa M et al (1998) Development of ELISA for the detection of Ralstonia solanacearum in tomato its application in seed health testing. World J Microbiol Biotechnol 14:697. https://doi.org/10.1023/A:1008892400077
19. Engvall E, Perlmann P (1972) Enzyme-linked immunosorbent assay, Elisa. 3. Quantitation of specific antibodies by enzyme-labeled anti-immunoglobulin in antigen-coated tubes. J Immunol 109:129–135
20. Andreasson U, Perret-Liaudet A, van Waalwijk van Doorn LJC et al (2015) A practical guide to immunoassay method validation. Front Neurol 6:179. https://doi.org/10.3389/fneur.2015.00179
21. Kelman A (1954) The relationship of pathogenicity in Pseudomonas solanacearum to colony appearance on a tetrazolium medium. Phytopathology 44:693–695
22. Granada G, Sequiera L (1983) A new selective medium for Pseudomonas solanacearum. Plant Dis 1084–1088. https://doi.org/10.1094/PD-67-1084
23. Elphinstone JG, Henessey J, Wilson J, Stead D (1996) Sensitivity of different methods for the detection of Pseudomonas solanacearum in potato tuber extracts. EPPO Bull 26:663–678. https://doi.org/10.1111/j.1365-2338.1996.tb01511.x
24. Kinyua Z, Miller S, Chin A, Subedi N (2014) Bacterial wilt disease Ralstonia solanacearum. Standard operating procedure for use in diagnostic laboratories. Version: EA-SOP- RS1. The international plant diagnostic network. https://ipmil.cired.vt.edu/wp-content/uploads/2014/06/SOP
25. Bellstedt DU (2009) Enzyme-linked immunosorbent assay detection of Ralstonia solanacearum in potatoes: the South African experience. In: Burns R (ed) Plant pathology. Methods in molecular biology (methods and protocols), vol 508. Humana Press, Totowa, NJ. https://doi.org/10.1007/978-1-59745-062-1_5
26. Paret ML, De Silva AS, Alvarez AM (2009) Bioindicators for Ralstonia solanacearum race 4: plants in the Zingiberaceae and Costaceae families. Australas Plant Pathol 38:6–12. https://doi.org/10.1071/AP08069
27. Suryadi Y (2009) Development of antibody to Ralstonia solanacearum and its application for detection of bacterial wilt. Biotropia 16:79–87. https://doi.org/10.11598/btb.2009.16.2.58

Chapter 2

Diagnosis of Tropical Fascioliasis Using Specific Antigen Detection in Feces of Infected Human Being or Animal Species

Neelima Gupta, Said I. Shalaby, Dileep K. Gupta, and Mona A. Awad

Abstract

Tropical fascioliasis is considered to be a significant factor in limiting livestock production. The development of sustained strategies for accurate early diagnosis is essential for controlling *Fasciola gigantica* infection. Diagnosis of *Fasciola* infection is usually achieved by identifying flukes' eggs in feces. This is time-consuming and of low accuracy. Moreover, by the time eggs are produced, damage to liver parenchyma may be severe. In an effort to develop more sensitive diagnostic methods, two types of immunoassays have been investigated. One depends on detection of antibodies in serum of infected animals which can detect infection from 2 to 4 weeks after infection, and the other, sedimentation technique, can detect eggs after 10–14 weeks postinfection. The second approach relies on detection of circulating parasitic antigen using ELISA, being detected 4–6 weeks of infection and decreases once worms enter bile ducts. However, serum testing of large herds may be difficult and fecal antigens can solve the problem. For the detection of anti-*Fasciola* antibodies in serum, sensitivity and specificity depend on the antigen used including *Fasciola* crude antigen, egg antigen, excretory secretory (ES) antigen, and coproantigen after several methods of purification. However, the specificity and sensitivity for diagnosis need some enhancement. On the other hand, the latter antigen (coproantigen), which was extracted from feces of infected animals, offers potential advantages on serological assays and depends on identification in stools, which being more accessible than serum. It detects only recent infection and is not time-consuming. It is stable at 4 °C. The amount of antigen can reflect worm numbers. A complete description of well-tested techniques is given.

Keywords *Fasciola gigantica*, Diagnosis, Coproantigen, Egg antigen, Excretory secretory (ES) antigen, ELISA

1 Introduction

Tropical fascioliasis is caused by infection with *Fasciola gigantica*. It is regarded as one of the most important helminthic infections in ruminants all over the world. Since the disease limits livestock production, the development of sustainable strategies for accurate early diagnosis is essential for controlling *F. gigantica* infection [1].

Diagnosis is usually achieved by finding eggs using fecal sedimentation method. The latter is time consuming and of low

Neelima Gupta and Varsha Gupta (eds.), *Experimental Protocols in Biotechnology*, Springer Protocols Handbooks, https://doi.org/10.1007/978-1-0716-0607-0_2,

accuracy. Immunoassays can help in solving this problem including antibody based serological tests and detection of circulating parasitic antigens being detected b ELISA [2]. ES antigens were described to be the best for immunodiagnosis of fascioliasis in cattle and sheep [3]. Antibodies were used against 88 kD, and it was shown that circulating antigen was detectable as early as second and third weeks after infection [4].

The sensitivity of egg antigen was 40% using ELISA, while its specificity was 95% [5]. Diagnostic coproantigens were detected in bile and feces from *F. gigantica* infected cattle [6] and 4 monoclonal antibodies were produced, capable of binding a low molecular weight band of *F. hepatica* ES antigen.

Diagnostic monoclonal antibody-based capture ELISA were evaluated for detection of 26–28 kD *F. hepatica* coproantigen in cattle and infection could be detected as early as 6 weeks [7].

The specificity of ELISA using ES antigen with optical density more than 0.38 allowed the differentiation among fascioliasis, schistosomiasis, and other human parasitic infections [8]. Some of parasites may cross react with crude *Fasciola* antigens in serological diagnosis [9]. These diseases include hydatidosis, amoebic liver abscess, heterophyiasis, and trichinosis. The authors added that partial purification of *Fasciola* antigen is a suitable method to avoid cross-reactivity.

Western blot assay technique with hyperimmune serum obtained from ES antigen of adult *F. hepatica* was used and antigens of possible diagnostic interests in patients' feces were found to have a molecular weight of 14, 19, 20, 23, and 51 kDa [10]. The authors used affinity chromatography to purify antigens in feces using ES 78 monoclonal antibody bound to CNBr-activated Sepharose 4B and suggested that polypeptides could be important markers for acute and chronic fascioliasis. Coproantigens have specific diagnostic ability, can be directly assessed for *Fasciola* infection, and have low numbers of cross-reaction proteins reflecting high specificity [11, 12]. Thus, some *Fasciola* antigens (coproantigen, ES antigen, and egg antigens) can be of possible diagnostic value and can be used against their related hyperimmune sera.

2 Technique

Copromicroscopical examination of animals, as well as blood samples for serum separation can be done whenever possible. Collected fecal samples are examined directly using Fluke Finder and Floatation techniques. Coproantigen can be prepared from feces of positively *Fasciola gigantica* infected cattle from which ES and Egg antigens are prepared. This can serve as a reference for evaluation of the diagnostic value of tested antigens.

Blood samples are used for separation of known *Fasciola*-infected serum samples. Serum is labeled and stored at −20 °C till use in immunological tests. The techniques indirect enzyme linked immunosorbent assay (ELISA) and monoclonal antibody sandwich ELISA are discussed.

3 Materials

3.1 Fecal Sampling

1. Animals with ear tag numbers.
2. Gloves and splash glasses to be worn when handling animals.
3. Plastic or glass containers for collecting fecal samples.
4. Fecal samples are directly examined using Fluke Finder and concentration flotation technique.

3.1.1 Fluke Finder Technique

1. Fecal samples.
2. Two successive sieve systems, the upper one of 400 μm/pore which permits the passage of eggs and fine particles, the second of 100 μm/pore which prevents eggs from passing.
3. Petri dishes.
4. Methylene blue dropper.
5. Stereo binocular microscope.

3.1.2 Concentration Flotation Technique

1. Saturated sodium chloride solution.
2. Glass slides.
3. Glass covers.
4. Glass test tubes.
5. Centrifuge.
6. Stereo binocular microscope.

3.2 Fasciola gigantica Excretory–Secretory (ES) Antigen Preparation

1. Bovine livers.
2. 0.01 M PBS, pH 7.4 (molar phosphate buffer saline).
3. Incubator.
4. High-speed centrifuge (12,000 rpm).
5. Deep Freezer (−70 °C).

3.3 Fasciola gigantica Egg Antigen Preparation

1. Bile of infected slaughtered animals.
2. Distilled water.
3. Penicillin G (500 IU/mL).
4. Gentamycin (50 mg/mL).
5. Sonicator.
6. Sonifier cell disruptor.

7. High-speed centrifuge (10,000 rpm).
8. Deep freezer (−70 °C).

3.4 *Fasciola gigantica* Coproantigen Preparation

1. Fecal samples.
2. Distilled water.
3. Sonicator.
4. Sonifier cell disruptor.
5. High-speed centrifuge (3000 rpm).
6. Dialysis tubes.
7. 36 kDa polyvinylpyrrolidone
8. Deep freezer (−70 °C).

3.5 Blood Sampling

1. Animals.
2. Gloves and splash glasses to be worn when handling animals.
3. 5 ml syringe or vacutainer needles.
4. Sterile needles.
5. Test tubes or small bottles for receiving blood.
6. Disinfectant and cotton for cleaning up after blood sampling.

3.6 Separation of Serum

1. Serum is separated after centrifugation of blood samples.

3.7 Indirect Enzyme-Linked Immunosorbent Assay (ELISA)

1. The test is carried out as described by Oldham [13].
2. Materials and reagents include the following:
3. 96-well flat bottom microtiter plates (Linbro, Flow laboratories, Connecticut, USA).
   (a) 0.01 M phosphate buffer saline solution, pH 7.4 (PBS):

KH_2PO_4	0.19 g
Na_2HPO_4	1.58 g
NaCl	8.00 g
Dist. Water	to 1000 mL

   (b) Washing buffer 0.01 M PBS, pH 7.4 containing 0.05% Tween 20 (PBS-T).
   (c) Coating buffer, 0.05 M carbonate buffer, pH 9.6.

N_2CO_3	1.59 g
$NaHCO_3$	2.93 g
Dist. Water to	1000 mL

Mix well and adjust pH at 9.6.

(d) Blocking buffer, 0.1% bovine serum albumin (BSA) in coating buffer.

(e) Serum diluents, PBS-T.

(f) Conjugate, horseradish peroxidase conjugated goat anti-rabbit IgG (whole molecule) (Sigma Immunochemicals). The conjugates are used at a concentration of 1:2000.

(g) Citrate/Phosphate buffer, pH 5.0.

A	Citric acid	2.10 g/100 mL Dist. water
B	$Na_2HPO_4 \cdot 12H_2O$	7.16 g/100 mL Dist. water
24.3 mL from A + 25.7 mL from B + 50 mL Dist. water		

(h) Substrate buffer, citrate /phosphate, pH 5.0 with 3.9 mM H_2O_2.

(i) Substrate, ortho-phenylenediamine (Sigma Immunochemicals) at a concentration of 340 μg/mL substrate buffer.

(j) Sera from rabbits experimentally infected with *Fasciola gigantica* (nonspecific infection), reference rabbit hyper-immune sera and sera from noninfected rabbits (as control).

(k) Antigens, *F. gigantica* coproantigen, ES antigen, and egg antigen, are used at an optimal concentration of 4 μg/mL coating buffer [13].

3.8 Monoclonal Antibody Sandwich ELISA

1. Antigen of the tested sample is captured between 2 antibodies.
2. The first one is specific monoclonal antibody used in coating ELISA plate.
3. The other one is laboratory prepared reference hyperimmune sera of other hosts [7].

4 Methods

4.1 Collection of Fecal Samples

1. Fecal samples are collected by per rectal collection.

4.2 Coproexamination

1. Fecal samples are examined directly for determining the parasitic status of each animal.
2. Direct examination of fecal samples is done using Fluke Finder and Floatation technique.

4.2.1 Fluke Finder Technique

1. Two grams of feces are applied on 2 successive sieve systems the upper one of 400 μm/pore which permits the passage of eggs and fine particles. The second of 100 μm/pore which prevents eggs from passing.
2. The content over the second sieve is transferred to a small petri dish using tap water.
3. One drop of methylene blue is then added.
4. Each sample is examined two times [14].

4.2.2 Concentration Flotation Technique

1. Two grams feces is added to 10 mL of the floatation solution (saturated sodium chloride solution).
2. Mix well and then power to glass test tube till full.
3. Place a glass cover on the top of the test tube.
4. Centrifuge the test tube at 1599 rpm for 5 min.
5. Remove the cover vertically and put on a slide and examine under the microscope [15].

4.3 Preparation of the Tested Antigens

4.3.1 Fasciola gigantica *Excretory–Secretory (ES) Antigen Preparation*

1. Worms are obtained from bovine livers.
2. Worms are washed in 0.01 M PBS, pH 7.4 for 3–5 times.
3. Worms are incubated for 3 h at 37 °C (1 worm/5 mL in 0.01 M PBS, pH 7.4).
4. The supernatant fluid is collected and subjected to high speed centrifugation (12,000 rpm) for 1 h at 4 °C.
5. The supernatant is separated and designated as ES antigen [3].
6. The protein content is measured [16].
7. The antigen is aliquoted and stored at −70 °C until use.

4.3.2 Fasciola gigantica *Egg Antigen Preparation*

1. Eggs are collected from the bile of infected slaughtered animals by sedimentation method using several changes of tap water.
2. Eggs are washed several times by distilled water containing penicillin G (500 IU/mL) and gentamycin (50 mg/mL).
3. Eggs are collected after sedimentation for 30 min.
4. Eggs are sonicated with 0.01 M PBS, pH 7.4 for 10 min under 150-W interrupted pulse output at 50% power cycle using a sonifier cell disruptor.
5. Sonicated eggs are subjected to a high-speed centrifugation (10,000 rpm) for 1 h at 4 °C.
6. Supernatant is separated as egg antigen after protein content is measured [16].
7. The antigen is aliquoted and stored at −70 °C until used [6].

4.3.3 *Fasciola gigantica* Coproantigen Preparation

1. Fecal samples of each parasitic stage can be used for preparation of *F. gigantica* coproantigen.
2. Five gram of sample is mixed separately in equal amount of distilled water and sonicated for 5 min under 150-W interrupted pulse output at 50% power cycle using a sonifier cell disruptor.
3. Fecal suspensions are centrifuged at 3000 rpm for 15 min.
4. The supernatant is dialyzed in 6–8 kDa dialysis tubes overnight at 4 °C against 4 M urea.
5. Fecal supernatants are then centrifuged to a fifth the original volume by absorption against 36 kDa polyvinylpyrrolidone.
6. The protein content is measured [16] and stored at −20 °C until use [6].

4.4 Sampling and Analysis

4.4.1 Method of Blood Sampling

1. Blood samples are taken from the jugular vein.

4.4.2 Serum Separation from Blood

1. Blood samples are used for separation of known serum samples.
2. Serum samples are labeled and stored at −20 °C till used in immunological tests.

4.4.3 Indirect Enzyme-Linked Immunosorbent Assay (ELISA)

1. It is used for current determination of circulating antibodies.
2. Antigens are diluted in coating buffer at their optimal dilution (4 μg/mL coating buffer).
3. Each well is filled with 200 μL of the corresponding antigen concentration and then the plates are incubated overnight at 4 °C.
4. Plates are washed three times with PBS-T 0.05% to get rid of excess unbound antigen and the remaining free binding sites are then blocked with BSA for blocking buffer (200 μL/well) and kept for 1 h.
5. Plates are washed three times with PBS-T 0.05%.
6. The sera are added to the plates (100 μL/well) and incubated at 37 °C for 90 min.
7. Plates are washed three times with PBS-T 0.05% and 100uL/well of the conjugate (1:2000 dilution in PBS-T 0.05%) are added to all wells and incubated at 37 °C for 1 h.
8. After incubation, the plates are washed five times with PBS-T and twice with substrate buffer; substrate is added at 100 μL/well and incubated for 5 min at room temperature.
9. The reaction with yellowish coloration is stopped by adding 100 μL/well of 1 M H_2SO_4 and is read using ELISA reader (Titerteck Multiscan—Dynatech Laboratories) at 490 nm.

10. A positive serum sample is defined as having an optical density (OD) value greater than the mean OD value for the noninfected controls plus two times the Standard Deviation (Cut off value) [15].

4.4.4 Monoclonal Antibody Sandwich ELISA

1. Used for capturing antigens in infested samples.
2. Flat-bottom 96-well microtiter plate is sensitized overnight at 4 °C, with 1.5 μg of saturated ammonium sulfate–purified monoclonal antibody F10/well, diluted in 0.05 M carbonate buffer, pH 9.6.
3. Plates are washed six times with PBS-T 0.05%.
4. The unbound sites in the wells are blocked for 2 h at room temperature with 200 μL/well of blocking buffer (3$ BSA/PBS with 0.02% sodium azide).
5. After incubation, the plates are washed six times with PBS-T 0.05%.
6. 100 μL/well of fecal supernatants diluted 1:1 in blocking buffer are added and incubated for 2 h at room temperature.
7. After incubation, the wells are washed as before and then rabbit hyperimmune sera (20 μg/mL) in blocking buffer are added at 100 μL/well and incubated at room temperature for 2 h.
8. Plates are washed six times as before and then 100 μL of peroxidase-conjugated goat anti-rabbit Ig G, diluted 1:1000 with blocking buffer is added to each well.
9. After incubation for 1 h, wells are washed six times as above and 100 μL of *o*-phenylenediamine substrate is added to each well.
10. The reaction is allowed to proceed for 15 min at room temperature in the dark and is stopped by dilution of 100 μL/well of 1 M H_2SO_4.
11. Color change is measured at 490 nm using ELISA-reader.
12. A positive fecal specimen is defined as having an OD value greater than the mean OD for noninfected controls plus 2 ties SD (cut off value).
13. Antigen determination is performed in duplicate and the results are expressed as the mean absorbance for each determination.
14. Antigen detection assay is evaluated in terms of its sensitivity (percentage of positive results among the total number of coprologically *Fasciola* positive animals) and Specificity (Percentage of negative results among total number of coprologically *Fasciola* negative animals) [7].

5 Notes

1. Wear gloves at all times when handling blood collection equipment and blood. This will help to reduce the risk of contamination [1].
2. Carefully remove the needle and empty all blood immediately into the appropriate micro centrifuge tube. If the needle is kept in place, the red blood cells may be lysed [1].
3. Never attempt to resheath needles. Always discard them into a sharps container immediately after use [1].
4. Ensure you are working in a clean environment. Allow sufficient workplace for obtaining blood samples without contamination [1].

References

1. Shalaby HA (2002) Evaluation of antigenicity and diagnostic value of some *Fasciola gigantica* antigens. Ph.D. thesis, Faculty of Veterinary Medicine, Cairo University, Cairo, Egypt
2. Knobloch J (1985) Human fascioliasis in Cajamarca/Peru. II—Humoral antibody response and antigenaemia. Trop Med Parasitol 36:91–93
3. Rivera-Marrero C, Santiago N, Hillyer G (1988) Evaluation of immunodiagnostic antigens in excretory secretory products of *Fasciola hepatica*. J Parasitol 74:646–652
4. Fagbemi B, Aderibigbe O, Guobadia E (1997) The use of monoclonal antibodies for immunodiagnosis of *Fasciola gigantica* infection in cattle. Vet Parasitol 69:231–240
5. Amer A (1996) The evaluation of different developmental stages of *Fasciola gigantica* antigens in the serodiagnosis of fascioliasis. M. Sc. thesis, Faculty of Science, Cairo University, Cairo, Egypt
6. El-Bahy M, Malone J, Todd W, Schnorr A (1992) Detection of stable diagnostic antigen from bile and faeces of *Fasciola hepatica* infected cattle. Vet Parasitol 45:157–167
7. Abdel-Rahman S, O'Reilly K, Malone J (1998) Evaluation of a diagnostic monoclonal antibody-based capture enzyme-linked Immunosorbent assay for detection of 26–28 kD *Fasciola gigantica* coproantigen in cattle. Amer J Vet Res 59:533–537
8. Espino A, Dumenigo B, Fernandez R, Finlay C (1987) Immunodiagnosis of human fascioliasis by ELISA using ES products. Amer J Trop Med Hyg 37:605–608
9. Khalil H, Abdel-Aal T, El-Zaat E (1990) Specificity of crude and purified Fasciola antigens in immunodiagnosis of human fascioliasis. J Egypt Soc Parasitol 20:87–94
10. Espino A, Bzrges A, Dumcnigo B (2000) Faecal antigens of *Fasciola gigantica* potentially useful in diagnosis of fascioliasis. Rev Panam Salud Publica 7:225–239
11. Shalaby SI, El-Bahy M, Hassan A, Shalaby H, Gupta N (2013) Diagnostic value of some *Fasciola gigantica* antigens. J Parasit Dis 39 (3):479–486. https://doi.org/10.1007/s12639-013-0379-1
12. Shalaby SI, El-Bahy M, Shalaby SIA, Shalaby H, Gupta N, Gupta DK (2014) Detection of coproantigens by sandwich ELISA in rabbits experimentally infected with *Fasciola gigantica*. Iran J Parasitol 9(3):374–381
13. Oldham G (1983) Antibodies to *Fasciola gigantica* antigens during experimental infections in cattle measured by ELISA. Vet Parasitol 13:151–158
14. Welch S, Malone J, Geaghan H (1987) Herd evaluation of *Fasciola hepatica* infection in Louisiana cattle by an ELISA. Am J Vet Res 48:345–347
15. Soulsby E (1982) Helminths, arthropods and protozoa of domesticated animals, 6th edn. Blaire Tindall and Cassell Ltd, London
16. Lowry O, Rosenbrough N, Farr A, Randall R (1951) Protein measurement with Folin-phenol reagent. J Biol Chem 193:265–275

Chapter 3

CMV Promoter-Driven Expression and Visualization of Tagged Proteins in Live and Fixed Zebrafish Embryonic Epidermis

Kirti Gupta and Mahendra Sonawane

Abstract

In the recent past, zebrafish has emerged as a highly useful vertebrate model for biomedical research. Owing to its easy handling, suitability for high-throughput genetic and chemical screens, and tractability by high-resolution microscopy, it is possible to study molecular mechanisms of vertebrate development and disease biology using zebrafish. This chapter introduces zebrafish epidermis as a model for studying cell biology of epithelial tissues in vivo and describes the technique of protein expression using plasmid vectors having cytomegalovirus (CMV) promoter. It details protocols for microinjection of plasmids into fertilized zebrafish oocytes, screening of the epidermal clones expressing the tagged protein, immunostaining, and mounting of embryos for both fixed and live imaging by confocal microscopy. This collection of protocols allows for analysis of localization of a wide range of proteins including those involved in intracellular transport, cell polarity, cell adhesion, and so on, during early developmental stages of zebrafish embryos and larvae.

Keywords Zebrafish, Microinjections, CMV promoter–based expression, Live imaging, Confocal microscopy

1 Introduction

Zebrafish embryonic and larval epidermis is progressively gaining importance as a model to investigate molecular and cellular mechanisms that are essential for epidermis development. The embryonic epidermis is a bilayered epithelial tissue having an outer peridermal layer and the underlying basal epidermis. This bilayered epithelial organization becomes apparent at around 18–24 h postfertilization (hpf) and can be observed till 15 days postfertilization (dpf) [1–3]. The epidermis subsequently undergoes further stratification [1]. Being the outermost tissue, epidermis is easily accessible for morphological observations under various genetic and chemical perturbations, and allows for high-resolution imaging to analyze phenotypes at the cellular level. Many

Neelima Gupta and Varsha Gupta (eds.), *Experimental Protocols in Biotechnology*, Springer Protocols Handbooks, https://doi.org/10.1007/978-1-0716-0607-0_3,

labs have used zebrafish as a model to address key questions in the field of epidermal biology concerning epithelial stratification, tissue polarity, cell differentiation, membrane dynamics, cell shape and size regulation, and tissue mechanics [2–11]. Several reagents—in terms of antibodies and transgenic lines driving expression of reporter genes such as GFP in the epidermis—have been used to address these questions [3, 5, 12, 13]. In addition, transgenic lines expressing epidermal proteins tagged with fluorescent markers are also available. Such lines yield useful insights into dynamics of live processes that cannot be obtained using antibodies. Unfortunately, such GFP/RFP fusion protein lines allowing for live imaging at subcellular resolution in the epidermis are rather limited [6, 14–16]. Repository of these transgenic lines and antibodies is available at https://zfin.org/. Although highly useful, this limited repertoire does not fulfill the growing requirements of the field. Besides, the process of generating transgenic lines is time and resource intensive, and even if lines are available, they are not easily accessible all over the world due to import/export related restrictions.

Microinjection of plasmid carrying the gene of interest is a useful and effective method for gene expression studies in zebrafish [7–9]. It allows for rapid assessment of localization and dynamics of the protein of interest as well as cellular compartments to which the protein localizes, when tagged with a fluorescent protein, such as GFP. CMV 1E94 promoter-based mammalian expression vectors pCS, pCS2 MT, pCS2+, pCS2P+, and pCS2+8 show strong ubiquitous expression of cloned genes [17–19]. The presence of SV40 polyadenylation signal in these vectors facilitates polyA tailing stabilizing the transcripts. Additionally, pCS2 plasmids have bacteriophage promoters such as SP6, T7, and T3 flanking the polylinker region, which allow for mRNA synthesis of the cloned gene by in vitro transcription. There are numerous vector backbones available as pCS2+8N/C-FP that allow for addition of fluorescent protein (FP) tag on either N or C terminal of the protein of interest.

The enveloping layer (EVL) or peridermal expression of plasmid encoded tagged proteins begins at around 8–10 hpf [8] and lasts until 2–3 dpf. Being a relatively fast approach, plasmid-based expression can be useful for quick screening of multiple proteins to analyze their intracellular localization. To achieve faithful localization shown by endogenous proteins, and to avoid overexpression artifacts, the amount of the plasmid injected needs to be carefully calibrated. Additionally, plasmid microinjections can also be used to achieve intentional overexpression to analyze gain of function phenotypes at the cellular level.

2 Technique

In this chapter, we have detailed the use of Cytomegalovirus (CMV) promoter-driven expression of fusion proteins as a method for in situ imaging in the epidermis of both live and fixed zebrafish embryos. The chapter presents protocols for plasmid selection and microinjection, screening of embryos having the clones, immunostaining, and imaging of zebrafish embryos to study localization and dynamics of proteins of interest encoded by plasmid vectors. The protocols compiled in the chapter are derived from the existing protocols, and standardized specifically for the epidermal tissue.

3 Material

3.1 Plasmid Preparation for Microinjections

Plasmid preparation kit, nuclease-free water (NFW), phenol red solution (0.5%, Sigma).

3.2 Microinjection of the plasmid at the one-cell stage in the zebrafish embryo (20)

Zebrafish lines, plasmids, plastic microinjection mold, agarose, E3 buffer (5 mM NaCl, 0.17 mM KCl, 0.33 mM $CaCl_2$, 0.33 mM $MgSO_4$), glass capillary needles, micropipette, microloader tips, fine forceps, micrometer, micromanipulator setup.

3.3 Screening and Fixation of the Embryos

Tricaine (Sigma-Aldrich) (0.168 mg/mL in E3 buffer), fine forceps, low melting temperature agarose, stereo microscope with a fluorescence module, E3 buffer, 90 mm petri dishes, 2 mL microfuge tubes, Pasteur pipettes, phosphate buffered saline (PBS; 0.8% (w/v) NaCl, 0.02% (w/v) KCl, 0.02 M phosphate buffer, pH 7.4), 4% paraformaldehyde (PFA) in PBS.

3.4 Immunostaining

PBS (pH 7.4), Pasteur pipettes, rocker and rotor instruments, PBT (0.8% Triton X-100 in PBS), Molecular biology grade methanol, normal goat serum (NGS), primary and secondary antibodies, aluminium foil, molecular biology grade glycerol solution.

3.5 Whole-Embryo Mounting for Imaging Epidermis

E3 buffer, Tricaine (Sigma-Aldrich) (0.168 mg/mL in E3 buffer), fine forceps, low melting temperature agarose, glass bottom petri plates, tissue paper, 11 no. scalpel blades, 18 mm coverslips (1 mm thick), glass slides, dissection microscope.

3.6 Confocal Microscopy

Laser scanning confocal microscope (e.g., Zeiss 510, Zeiss 710) with lasers for exciting different fluorophores, required objectives (e.g., 10×, 40× oil, 63× oil) and metal halide lamp.

4 Methods

4.1 Plasmid Preparation for Microinjections

A repository of pCS plasmids is available online (https://www.addgene.org/), and if plasmids encoding the protein of interest are available in the library, they can be ordered from Addgene. Alternatively, the gene of interest can be cloned in pCS2+/pCS2+8-FP vectors using any standard cloning protocol. The expression, localization, and function of the recombinant fluorescent protein might vary with the nature and location (C- or N-terminal of the protein) of the fluorescent tag. To circumvent these issues, it is important to perform domain analysis of the protein of interest and make a reasonable choice between C- terminal or N- terminal tags. Additionally, cloning should be done in a directional manner in the multiple cloning site (MCS), such that the promoter (sp6/T7/T3) should be upstream of the gene of interest for mRNA synthesis, if required.

These plasmids are transfected and isolated by standard plasmid isolation protocol using kits; however, final elution is done using nuclease-free water (NFW) rather than elution buffers. Purity of plasmid should be high ($A_{260/280} > 1.8$ and $A_{260/230} > 2$) to minimize toxicity upon injections. The appropriate plasmid concentration for the microinjections could be determined as follows-

1. Standardize the concentration required for the expression by testing different dilutions of the plasmid. The concentration range of 20 ng/μL to 120 ng/μL works the best for most of the plasmids. Microinject (see Sect. 4.2) different concentrations of the plasmid, 20, 40, 80, 100, 120 ng/μL, diluted in NFW and 10% phenol red solution, for standardization. The higher amount of plasmid may cause morphological defects as well as formation of aggregates of expressed protein inside the cells. Plasmid toxicity may also result in extrusion of clones from the periderm.
2. For microinjection, select the concentrations at which no morphological defects are observed. Then select the lowest plasmid concentration that yields protein expression detectable by the confocal microscope.
3. Plasmids dissolved in NFW can be stored at 4 °C for a few weeks.

4.2 Microinjection of the Plasmid at One-Cell Stage in the Zebrafish Embryo [20]

1. Collect embryos, wash with E3 buffer, and arrange the embryos in the grooves made in 3% agarose in a petri plate using plastic microinjection mold (Fig. 1).
2. Load 2–5 μL of plasmid solution into a glass microcapillary needle using a pipette with microloader tips (*see* **Note 1**).

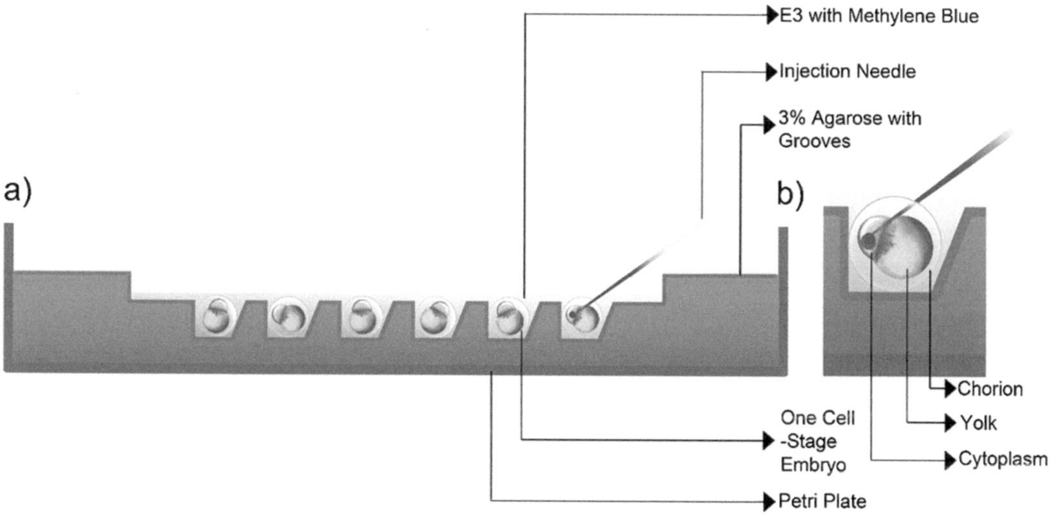

Fig. 1 Schematic representation of the process of microinjection at one-cell stage. A petri plate with 3% agarose with grooves is used to organize embryos at one-cell stage (**a**). The embryos remain submerged in sufficient amount of E3 during microinjections. Plasmids, dissolved in NFW and having phenol red as an indicator, are injected using a glass needle (**a**). High magnification (**b**) depicts the embryo details and the bolus microinjected in the cytoplasm

3. Cut the needle with the forceps from the tip such that bore size is not too big to injure the embryo yet not too thin to cause flexing during injections.
4. Measure the drop size by using a stage micrometer and adjust the drop size around 0.1 mm in diameter (0.5–1 nL) by regulating release pressure and time on the micromanipulator of the injection set-up.
5. Insert the needle into the chorion and bring the needle at 45° angle with the cell-yolk interface to pierce it and release the drop in the cytoplasm (*see* **Note 2**) (Fig. 1).
6. Collect the injected embryos into a petri dish by gently dislodging them from the mold by a jet of E3 buffer.
7. Remove dead, damaged, or deformed embryos by inspecting them under stereomicroscope after 3–4 h.

4.3 Screening and Fixation of the Embryos

Since the plasmid injection at one-cell stage gives rise to mosaic expression, it is important to screen for the embryos that have clones present in the periderm before proceeding to the imaging step. Depending upon the efficiency of microinjection and nature of the protein of interest, number of clones obtained may vary (*see* **Note 3**).

1. The embryos can be screened for the labeled clones in the periderm at or after 18 hpf.

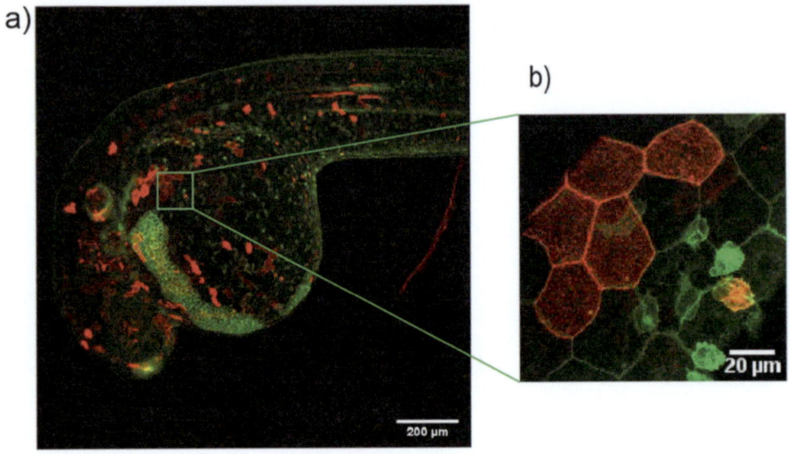

Fig. 2 Analysis of plasmid-injected embryos for epidermal clones expressing the tagged protein. Confocal image of a zebrafish embryo at 30 hpf shows clonal expression of LifeAct RFP (**a**) in the background of *Tg (cldnB:lynEGFP)* line. High-magnification image (**b**) of the region marked in "**a**" shows peridermal cells marked by LynGFP and clones that are additionally marked by LifeAct RFP

2. For screening the larvae beyond 48 hpf, use 0.04% Tricaine solution made in E3 buffer to anesthetize the larvae.
3. Use a stereomicroscope with high zoom and a fluorescence detection module. We use Zeiss SteREO Discovery for this purpose.
4. For proteins showing weaker expression or those with restricted localization within cells, for example on vesicles or cell organelles, examine carefully at a higher magnification on a stereomicroscope. The expression and localization may further be confirmed by performing confocal microscopy (Fig. 2) as discussed below (see Sect. 4.6).
5. Collect the embryos showing clones in the epidermis by a Pasteur pipette and transfer to a 90 mm petri dish.
6. For fixation, dechorionate embryos (if before 48 hpf) with fine forceps, transfer to a 2 mL microfuge tube along with E3 buffer (around 20–40 embryos can be fixed per tube), remove excess of E3 buffer and fix the embryos by adding 4% PFA (stored at 4 °C).
7. Incubate the embryos in PFA for 30 min at the room temperature followed by 12 h or overnight incubation at 4 °C before proceeding to immunostainings. Alternatively, fixation can be done at room temperature for 3–4 h.

4.4 Immunostaining

Immunostaining using antibodies against the fluorescent protein is useful for studying localization of the tagged protein with other known proteins or to a specific compartment, such as the nucleus or the Golgi (*see* **Note 4**). The protocol for immunostaining is discussed below.

1. On day 1, remove the 4% PFA fixative by giving three washes of 5 min each in ~2 mL PBS on a rocker.
2. Serially upgrade the embryos in methanol by giving 10 min washes on rocker in 30%, 50%, 70% methanol made in PBS, followed by 1 h incubation in 100% methanol. Although this step improves cell permeability, some antibodies do not work upon methanol post-fixation. Steps 2–4 should be skipped for such antibodies.
3. Samples upgraded in 100% methanol can be stored at −20°C for a few weeks or should be incubated for at least 3–4 h before proceeding to day 2 of IHC.
4. On day 2, serially downgrade the samples in 70%, 50%, and 30% methanol by giving 10 min wash each on rocker.
5. Give one 10 min wash of PBS on rocker followed by 4–5 washes of 10 min each in PBT.
6. Proceed with the blocking step with 10% NGS in PBT. Remove the PBT from previous wash from the tube and add 200 μL of blocking solution, making sure that all the embryos are submerged in the solution and leave on the rotor for 4 h.
7. Make appropriate primary antibody dilution in 1% NGS in PBT.
8. Remove the blocking solution from the tube and add 100 μL primary antibody solution. Make sure that all the embryos are submerged in the solution and leave overnight at 4 °C on a rotor for binding.
9. On day 3, remove the antibody solution and give five washes of 30 min each using PBT.
10. Prepare appropriate dilution of fluorescently conjugated secondary antibody in 1% NGS in PBT.
11. Remove PBT from the sample tubes as much as possible and add 100 μL secondary antibody solution. Wrap the tubes in aluminum foil to avoid photobleaching of the fluorophore and leave on a rotor at room temperature for 4 h.
12. Remove the secondary antibody solution and give four washes of 15 min each in PBT.
13. Postfix the sample in 4% PFA at room temperature for 30 min to cross-link the bound antibodies.
14. Remove PFA, wash the samples twice for 5 min each using PBS, and serially upgrade in 30%, 50%, and 70% glycerol made in PBS.

4.5 Whole-Embryo Mounting for Imaging Epidermis

4.5.1 Mounting of a Live Embryo

Live imaging allows for analysis of dynamic events such as transport of a protein and changes in an organelle or a cell. There are different ways for mounting embryos for live imaging such as mounting in E3 media, agarose, and methyl cellulose [20]. We have described agarose mounting, as it has been found to be suitable for both long- and short-term confocal imaging.

1. Anesthetize embryos with 0.04% Tricaine solution in E3 buffer.
2. Glass bottom petri dishes (35 mm) are used to mount the embryos in 0.5–1% low melting agarose (LMA). Heat LMA to 65 °C until it melts completely and allow it to stay at 35–40 °C.
3. Fill the glass bottom well in the petri dish with LMA and allow it to cool to room temperature and add roughly 1:10 volume of 0.4% Tricaine solution, to the agarose (*see* **Note 5**).
4. Pick up anesthetized embryos in a pipette and allow the embryos to settle at the tip of the pipette under gravity.
5. Gently transfer the embryo to the agarose well with minimal E3 buffer along with it.
6. Use blunt forceps or a needle to gently push the embryos down in to the agarose such that the head, yolk, flank, or any other region, which needs to be imaged, is pressed against the glass bottom with minimum agarose in between. This is a critical step since more agarose between the tissue and the glass bottom will not allow high magnification lens with limited working distance (63× oil, 100× oil) to focus on the tissue.
7. To image a specific region of the epidermis, position finely with needle or blunt forceps. For example, to image head epidermis, gently lift the tail with the forceps or needle such that the embryo stands on the widest part of the head against the glass slide (Fig. 3) (*see* **Note 6**).
8. Let the agarose cool down for 5–10 min (LMA solidifies below 25 °C), and add few drops of E3 at the top of the agarose or place an E3 soaked small piece of tissue paper at the rim inside the dish.
9. Cover the petri dish with the lid and proceed for confocal imaging on an inverted microscope.

4.5.2 Mounting of a Fixed Embryo

1. Fixed embryo or a part of it can be mounted in 70–80% of glycerol by placing it on a glass slide and positioning according to the site to be imaged.
2. For imaging the head epidermis, dissect out the embryo head by making a cut with a scalpel blade along the line joining the head and the yolk (Fig.4).

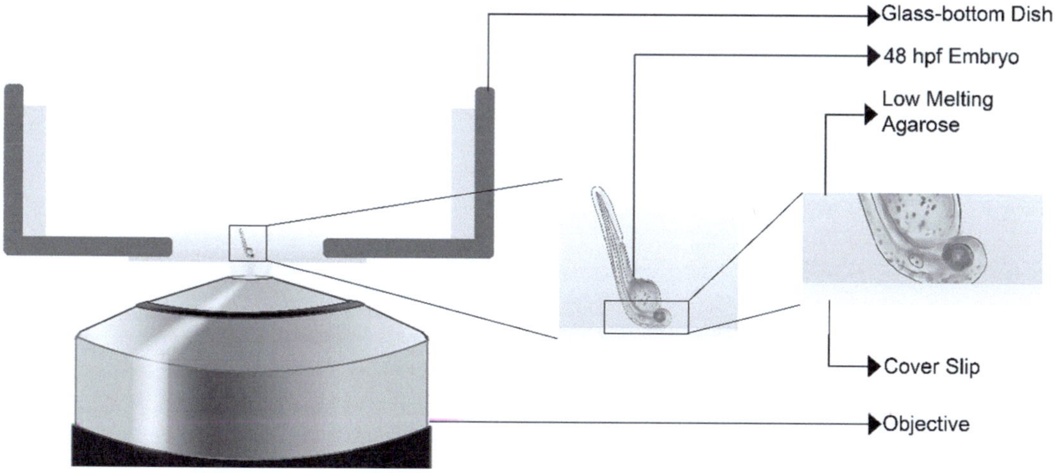

Fig. 3 Schematic showing mounting for live imaging of the head periderm in a 48-h-old embryo using inverted confocal microscope. An embryo having peridermal clones is mounted in low-melting agarose in an oblique fashion with its head pressing against the coverslip of the glass-bottom dish. The glass-bottom dish is placed on the microscope stage and imaging is performed using inverted confocal microscope

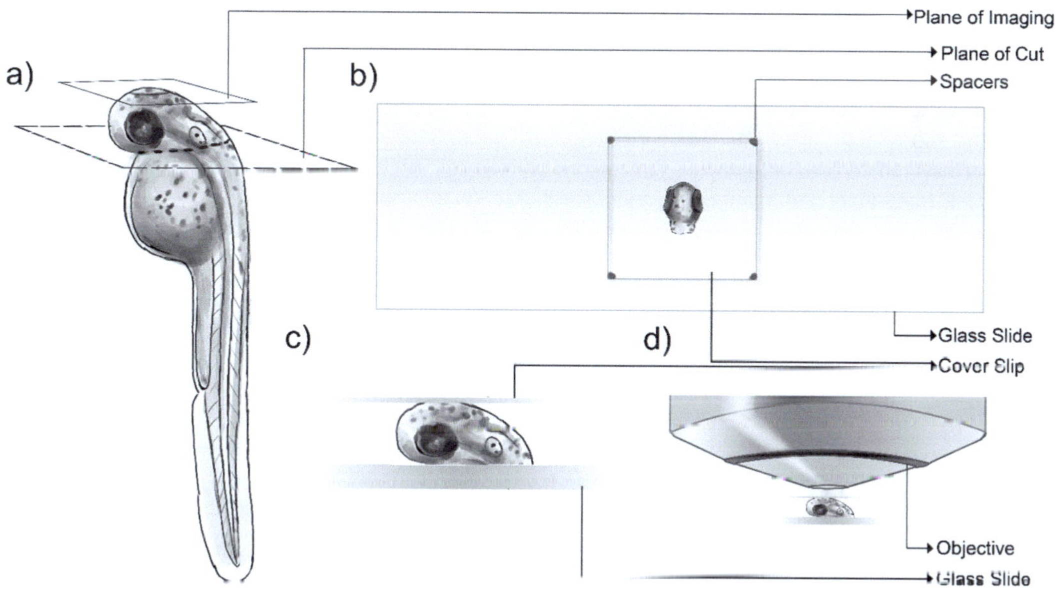

Fig. 4 Schematic showing dissection and mounting of a fixed embryo for imaging clones. A zebrafish embryo (**a**) showing the plane of cut (dotted line) for dissection of the head for mounting. The dissected head is placed on a glass slide in a drop of 70% glycerol with dorsal facing up (**b**, **c**). A coverslip is placed on the sample using plasticine at the corners as spacer (**b**) to avoid squashing of the head, and the sample is imaged on either an upright (**d**) or an inverted microscope

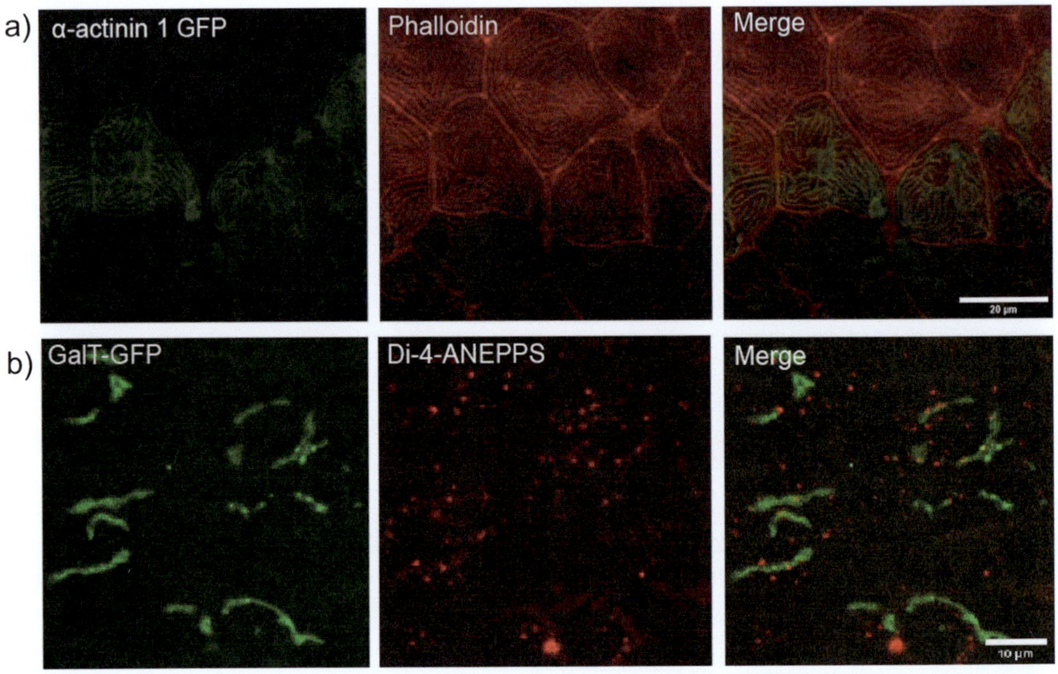

Fig. 5 Peridermal cells showing expression and localization of the expressed fusion proteins. (**a**) Confocal micrographs of a fixed clone of peridermal cells expressing plasmid-encoded α-actinin 1 conjugated with GFP. Note that α-actinin-1-GFP, which is an actin-binding protein, localizes to the microridges labeled using phalloidin rhodamine dye. (**b**) Live imaging of peridermal cells expressing GFP-tagged GalT (1,4-galactosyltransferase), an enzyme that marks trans-Golgi, and stained with Di-4-ANNEPS membrane dye to mark trans-Golgi and membrane-bound vesicles

3. Place the cut side of the head on the glass slide such that the epidermis on the middle of the head is facing up. Remove excess glycerol.
4. Put small bits of plasticine clay as spacers at the corners of an 18 mm coverslip, put the coverslip over the specimen and press gently while observing it under the microscope.

4.6 Confocal Microscopy Imaging

Epidermis, being the surface tissue, is easily accessible for imaging using a confocal microscope at high magnification and resolution to study intracellular structures and processes. We use Zeiss 710, 510 and Olympus fv1000, 1200 inverted microscopes with oil immersion lens for our imaging.

We have used the protocol described here to perform localization studies of two different fusion proteins, α-actinin-1-EGFP marking the microridges and GalT-GFP labeling trans-Golgi in the peridermal cells in fixed and live embryos, respectively (Fig. 5).

1. Switch on the microscope and initiate the lasers and the metal halide lamp at least 20 min before the usage.

2. Initiate the software that controls the confocal microscope and set the light path.
3. Secure the mounted embryo slide or glass bottom dish on the microscope stage with coverslip facing the lens.
4. Locate the embryo using lamp, and focus on the specimen with 10× lens without oil.
5. Once the embryo is located and is in focus, shift to the higher magnification oil immersion lens.
6. Visualize the tissue using high magnification lens under UV lamp to select the clone showing protein expression.
7. Shift to the laser scanning mode and do a quick scan on "live" mode to visualize the sample with laser illumination. Set the laser power, gain, and offset to ensure the coverage of maximum dynamic range for the system and to procure the best signal-to-noise ratio for the images (*see* **Note 7**).
8. Select the first and the last Z-sections. For peridermal cells, which spans around 4–6 μm in height, we find Z-sections of 0.379 μm to be optimal at 63×/1.4 oil objective.
9. Set the scanning speed and averaging according to the quality of the signal in the samples.
10. Set the zoom and pixel resolution considering the Nyquist criteria for the setup and resolution required for the experiment.
11. Screen the images for any large aggregates, which may be due to over expression of the protein, or any extruding cells from the periderm, which might be due to toxicity. Plasmid concentration should be reduced in such cases.

5 Notes

1. Microcapillary needles are made by pulling borosilicate glass capillaries of internal diameter 0.50 mm using needle puller by Sutter instrument p-97 (we use the following settings—heat = 800, pull = 63, vel = 60, del = 44).
2. We have observed that when injected inside the cytoplasm, the efficiency of plasmid expression is better.
3. In case of very low number of clones or if ubiquitous expression is required, mRNA of the protein can be synthesized by in vitro transcription and injected at one-cell stage.
4. For primary screening of protein expression, directly upgrade the PFA fixed embryos in glycerol after 3 PBS buffer washes, and proceed to microscopy.

5. Glass bottom dish (35 mm) are commercially available or can be made by cutting out ~14 mm diameter disk at the bottom and sticking a glass coverslip (1 mm thick) with the help of paraffin wax.
6. We mount the embryos to visualize head periderm, since cells are taller in this region. In addition, the dorsal part of the head is relatively flat which allows for better quality imaging. Flank and median fin fold can be visualized by placing the embryo laterally and gently pressing it against the glass bottom.
7. For imaging fixed samples, use same configurations as that for live imaging setup. Since signal is amplified by immunostaining, the laser power required for immunostained samples will be less.

Acknowledgments

We would like to thank Sumit Sen for the illustrations, Dr. Clyde Pinto for contributing to Fig. 5, Geetika Chouhan and Mandar Phatak for critically reading the manuscript, and Kalidas Kohale for fish maintenance. This work was supported by funding from TIFR-DAE (12P-121).

References

1. Le Guellec D, Morvan-Dubois G, Sire JY (2004) Skin development in bony fish with particular emphasis on collagen deposition in the dermis of the zebrafish. Int J Dev Biol 48:217–231
2. Sonawane M (2005) Zebrafish penner/lethal giant larvae 2 functions in hemidesmosome formation, maintenance of cellular morphology and growth regulation in the developing basal epidermis. Development 132:3255–3265
3. Lee RTH, Asharani PV, Carney TJ (2014) Basal keratinocytes contribute to all strata of the adult zebrafish epidermis. PLoS One 9(1): e84858
4. Sonawane M, Martin-Maischein H, Schwarz H et al (2009) Lgl2 and E-cadherin act antagonistically to regulate hemidesmosome formation during epidermal development in zebrafish. Development 136:1231–1240
5. Reischauer S, Levesque MP, Nüsslein-Volhard C et al (2009) Lgl_2 executes its function as a tumor suppressor by regulating ErbB signaling in the zebrafish epidermis. PLoS Genet 5(11): e1000720
6. Morris JL, Cross SJ, Lu Y et al (2018) Live imaging of collagen deposition during skin development and repair in a collagen I—GFP fusion transgenic zebrafish line. Dev Biol 441 (1):4–11
7. Raman R, Damle I, Rote R et al (2016) APKC regulates apical localization of Lgl to restrict elongation of microridges in developing zebrafish epidermis. Nat Commun 7:11643
8. Pinto CS, Khandekar A, Bhavna R et al (2019) Microridges are apical epithelial projections formed of F-actin networks that organize the glycan layer. Sci Rep 9:12191
9. Slanchev K, Carney TJ, Stemmler MP et al (2009) The epithelial cell adhesion molecule EpCAM is required for epithelial morphogenesis and integrity during zebrafish epiboly and skin development. PLoS Genet 5:e1000563
10. Webb AE, Driever W, Kimelman D (2008) Psoriasis regulates epidermal development in Zebrafish. Dev Dyn 237:1153–1164

11. Morita H, Grigolon S, Bock M et al (2017) The physical basis of coordinated tissue spreading in Zebrafish gastrulation. Dev Cell 40:354–366
12. Haas P, Gilmour D (2006) Chemokine signaling mediates self-organizing tissue migration in the zebrafish lateral line. Dev Cell 10:673–680
13. Eisenhoffer GT, Slattum G, Ruiz OE et al (2017) A toolbox to study epidermal cell types in zebrafish. J Cell Sci 130:269–277
14. Gong Z, Ju B, Wang X et al (2002) Green fluorescent protein expression in germ-line transmitted transgenic zebrafish under a stratified epithelial promoter from keratin8. Dev Dyn 223:204–215
15. Clark BS, Winter M, Cohen AR et al (2011) Generation of Rab-based transgenic lines for in vivo studies of endosome biology in zebrafish. Dev Dyn 240:245–265
16. Asakawa K, Kawakami K (2010) A transgenic zebrafish for monitoring in vivo microtubule structures. Dev Dyn 239:2695–2699
17. Miyoshi H, Blömer U, Takahashi M et al (1998) Development of a self-inactivating lentivirus vector. J Virol 72:8150–5157
18. Turner DL, Weintraub H (1994) Expression of achaete-scute homolog 3 in Xenopus embryos converts ectodermal cells to a neural fate. Genes Dev 8:1434–1447
19. Gökirmak T, Campanale JP, Shipp LE et al (2012) Localization and substrate selectivity of seaurchin multidrug (MDR) efflux transporters. J Biol Chem 287:43876–43883
20. Nusslein-Volhard C, Dahm R (2002) Zebrafish: a practical approach. Oxford University Press, New York

Chapter 4

Detection of Blood Parasites and Estimation of Hematological Indices in Fish

Neelima Gupta and Shahla Nigar

Abstract

Hematological parameters serve as excellent tools to monitor health status for diagnosis and prevention of diseases. Techniques in fish hematology are different from those of mammals due to the presence of nucleated erythrocytes in fishes. It is widely accepted that blood analysis will prove to be a valid diagnostic aid to promote aquaculture as it is predicted that by the year 2030, it will be the prime source of food. Popularly, the assessment of fish blood parameters has been performed manually, automatic blood cell count has also been performed in fish, but the techniques are not in much practice. A complete description of the techniques and methods employed for examination of fish blood for the identification of blood parasites and heme indices are described including the methods of blood sampling, isolation and processing of blood parasites, blood film preparation, and estimation of hematological parameters (TEC, TLC, DLC, Hb, PCV, ESR, MCV, MCH, MCHC). Blood collection and processing it for diagnostics need to be reliable, and the methods described herein include only well-tested and well-proven procedures and techniques for the determination of the different hematological parameters/indices.

Keywords Blood collection, Blood analysis, Blood film, Staining, Blood parasites, Hematological parameters

1 Introduction

Blood is a useful indicator of the physiological state of the animal. It has been used in the screening procedure to evaluate the general health, as an adjunct for the diagnosis of diseases to assess the body's ability to fight against infection and to evaluate the progress of diseases [1]. The values of the different hematological parameters are significantly influenced by endogenous and exogenous factors. Being nutritionally rich, blood serves as a favorable site for parasites. Once present, they may interfere with the blood physiology and terminate in diseased condition. Initiation of disease is a complex process involving the interactions between the parasite, host, and environment, creating a conducive condition for disease outbreaks. Some parasites spend most or all of their life cycle in the

Neelima Gupta and Varsha Gupta (eds.), *Experimental Protocols in Biotechnology*, Springer Protocols Handbooks, https://doi.org/10.1007/978-1-0716-0607-0_4,

bloodstream, parasites are known to inhabit all groups of vertebrates from fishes, to amphibia, reptiles, birds, and mammals [2]. The blood microniche has not escaped parasitism and has acted as a suitable environment of hemoparasitic invasion. "Parasitism in blood" has been a matter of curiosity and interest amongst parasitologists and hematologists all around the world [3]. The parasites include mastigophoreans *Trypanosoma*, the monoflagellates and *Trypanoplasma*, the biflagellates. These are extracellular blood parasites inhabiting the blood plasma and move actively in the blood by means of flagella. The sporozoans *Haemogregarina*, *Lankesterella*, *Hepatozoon*, *Plasmodium*, *Haemoproteus*, *Dactylosoma*, *Babesiosoma*, *Leucocytozoon*, *Myxobolus*, *Loma*, *Ichthyosporidium*, *Microsporidium*, *Pseudoloma*, *Spraguea*, *Microgemma*, and *Thelohania* also inhabit the blood extracellularly or intracellularly, and detection of all these parasites requires proper blood examination [4, 5]. Moreover, aquatic pollutants may alter fish hematology; therefore, toxicological studies are to be carried out to ascertain the effects of aquatic pollutants in the blood of fishes [6]. An alteration in hematological profile may cause biochemical dysfunction. The development of different techniques to allow multiple blood collection from fishes provides a useful tool to monitor the health of fish by knowing the normal value with respect to their responses to stress affecting body metabolism [7]. Although tremendous advances have been made in the field of hematological analyzers, examination of a well prepared, well stained blood smear still remains the cornerstone of diagnostic hematology. Even the most sophisticated hematology instruments are unable to consistently provide accurate differential cell counts, and no analyzer is capable of accurately identifying morphological changes, hemoparasites, neoplastic cells, and hematologic disorders etc.

Evaluating blood parameters involves peripheral blood smear examination, total erythrocyte count (RBC), total white blood cell count (WBC), hematocrit (PCV), hemoglobin concentration (Hb), erythrocyte sedimentation rate (ESR), and erythrocyte indices (MCV, MCH, and MCHC). The technique of blood sampling is a one-operator procedure not requiring the help of an assistant or any special equipment to restrain the fish. The operation of the protocol is unique since it permits the continuous collection of blood from the same experimental fish over a varied time course and reduces the need for a large number of replicate animals [8]. Numerous studies [9–11] have documented that changes in hematological indices depend on fish species, sexual maturity cycle, chemical pollutants like heavy metals [12], and so on. Thus, fish blood parameters act as a potential tool for identification of parasites, stress caused by environmental factors, and chemical intoxication.

2 Technique

This chapter deals with the techniques of blood sampling and examination specially for the identification of blood parasites of fish and techniques associated with examination of blood for hematological indices which may be useful for standardization of heme indices in different species of fish, for the detection and identification of blood parasites and to evaluate changes in the blood parameters due to intoxication, water pollution, and effects of parasite infection. The techniques of obtaining blood are described (from caudal vein, cardiac puncture, caudal severance, dorsal aorta, and kidney). The examination of blood deals with the isolation and processing of blood parasites by hanging drop preparations, and concentration of parasites by clotting blood and microcentrifugation. The methods of fixation and staining by stains (Giemsa, Leishman, Wright, Hemacolor, Giemsa azur eosin methylene blue, May–Grunwald–Giemsa, and Gomori trichrome staining) are described. The techniques of determination of blood parameters offer determination of RBC, WBC, hemoglobin, hematocrit, erythrocyte sedimentation rate, and derived hematological indices (MCV, MCH, MCHC).

3 Materials

3.1 Blood Sampling

1. Fish.
2. Buffered 0.2 mg/mL tricaine methane sulfonate (MS-222, Sigma), CIFECALM (2 drops in 1 L water) or similar appropriate anesthetic agent.
3. Gloves and splash glasses to be worn when handling fish or anesthesia water.
4. Scalpel or razor blade.
5. Cutting board or surface suitable for cutting.
6. 1 mL syringe or vacutainer needle (only for fish >300 g).
7. 1–5 mL heparinized syringe or vacutainer needle (for anesthetic fish).
8. Sterile needles (22 gauge, 1–1.5″ needle; however, size will depend on the weight of the fish).
9. Disinfectant for cleanup after blood sampling (70% isopropyl alcohol).
10. Soft sponge or mat to prevent mucous sloughing during blood sampling.
11. Dip net to be used for recovery container.
12. Heparinized pipette.

13. Vacutainer holder.
14. Heparin.
15. Microhematocrit tubules.
16. Saline solution.
17. Recovery tank or container.
18. Airstones, airline, and air pump.

3.2 Blood Film Preparation

1. Sterilized slides.
2. Sterilized cavity slides.
3. Slide box.
4. Coverslips.
5. Needle.
6. Methanol.
7. Dropper.
8. Cotton.
9. Staining racks/trough.
10. Coplin jars.
11. Volumetric flask or measuring cylinder.
12. Beakers.
13. Conc. H_2SO_4.
14. Potassium dichromate.
15. Cleansing solution.
    (a) Chromic acid or 70% alcohol.
    (b) 30% ether.
16. Glycerol.
17. Schaudinn's fluid.
18. Ethanol (Absolute).
19. Conc. HCl.
20. Phosphotungstic acid.
21. Chromotrope.
22. Fast green FCF.
23. Glacial acetic acid.
24. Distilled water.
25. Immersion oil.
26. Buffer tabs.
27. Giemsa stain.
28. Leishman stain.

29. JSB Stain (round bottom flask, distilled water, methylene blue (medicinal), H_2SO_4, $K_2Cr_2O_7$, $Na_2HPO_4{\cdot}2H_2O$, Eosin powder (Method of preparation given below).

    (a) JSB I: In 1 L round bottom flask, dissolve 500 mL of distilled water and 500 mg of methylene blue (medicinal). Add 1% of dilute H_2SO_4 dropwise and 3 cm^3 of $K_2Cr_2O_7$ and shake properly. A purple precipitate will be observed. 3.5 g of $Na_2HPO_4{\cdot}2H_2O$ is added and mixed thoroughly till the precipitate dissolves. The solution is boiled for 1 h at low temperature till it turns deep blue with a pinkish shade (end point). It is cooled, filtered and stored in a bottle and allowed to mature for 48 h. Filter the stain prior to use.
    (b) JSB II: 1 g of Eosin powder is mixed with 500 mL of distilled water. It is stored in bottle and filtered when used.

30. Buffer solution (pH 5.5–6.5) (Na_2 HPO_4, KH_2PO_4).

    (a) Na_2HPO_4—0.417 g, KH_2PO_4—0.752 g is mixed in 2000 mL of distilled water (pH 5.5–6.5).

31. Hemacolor® (Merck).

    (a) Fixing solution (Soln 1).
    (b) Two buffered staining solutions (Soln 2 and 3).
    (c) Buffered tablets.

32. Wright stain.

    (a) Absolute methanol.
    (b) Staining solution (Dissolve 0.3 g of Wright stain powder in 100 mL absolute methanol) and leave in a closed container at room temperature for 24 h. Filter before use.
    (c) Sorensen's buffer solution at pH 6.4.
        - Stock solution A: 0.1 M KH_2PO_4 (Dissolve 13.61 g KH_2PO_4 in 1000 mL of distilled water).
        - Stock solution B: 0.1 M Na_2HPO_4 (Dissolve 17.8 g $Na_2HPO_4{\cdot}2H_2O$ in 1000 mL of distilled water).
        - Mix stock solution A (70 mL) and stock solution B (30 mL) to give Sorensen's buffer solution (pH 6.47).

33. Giemsa Azur Eosin Methylene Blue.

    (a) Dissolve 1 g of Giemsa Azur eosin methylene blue in 54 mL of glycerol by heating. After cooling, add 84 mL methanol GR and filter the solution.

34. May–Grunwald–Giemsa Stain.
    (a) Absolute methanol.
    (b) Staining solution I: Leave 0.3 g May–Grunwald powder in 100 mL absolute methanol in closed container at room temperature for 24 h. Filter before use.
    (c) Staining solution II (Giemsa stain): Dissolve 1 g Giemsa stain powder in 66 mL glycerol and heat to 56 °C for 90–120 min. After addition of 66 mL absolute methanol and thorough mixing, leave the solution at room temperature in a closed container. Filter before use.
    (d) Buffer: Sorensen's buffer solution. The pH is 6.8 for the May–Grunwald–Giemsa stain instead of 6.4 as in the Wright stain. For pH 6.8, Stock solution A (50 mL) and B (50 mL) (prepared as given above for Wright stain), mixed to give Sorensen's buffer solution.
35. Heidenhain's iron hematoxylin stain.
36. Harris's hematoxylin.

Ammonium potassium alum	20.0 g
Hematoxylin	1.0 g
Mercuric oxide	0.5 g
Distilled water	100.0 mL
Absolute ethyl alcohol	10.0 mL

37. Xylene.
38. DPX or Canada Balsam.

3.3 Examination of Blood Parameters

1. Volumetric flask or measuring cylinder.
2. Beakers.
3. KH_2PO_4.
4. Na_2HPO_4.
5. pH meter.
6. Sterilized slides.
7. Slide box.
8. Stock Romanowsky stain like Giemsa stain.
9. Slide staining and drying rack.
10. Methanol.
11. Absolute alcohol.
12. Wash bottle.
13. Cotton.

14. Dropper.
15. Needle.
16. Forceps.
17. Coplin jars.
18. Cotton.
19. Pasteur's pipette (dropper with long nozzle).
20. EDTA vial.
21. Neubauer's hemocytometer.
22. Improved Neubauer chamber.
23. Hayem's solution (diluting fluid).
24. Turk's solution (diluting fluid).
25. EDTA solution (1.50 for 0.25 mg/mL of blood).
26. Sahli's hemoglobinometer.
27. Distilled water.
28. HCl 0.1 N.
29. Photoelectric colorimeter with green filter.
30. Drabkin's solution.

Potassium ferricyanide	200 mg
Potassium cyanide	50 mg
Potassium dihydrogen phosphate	140 mg
Nonionic detergent	1 mL
Distilled water	Make up to 1000 mL

31. Heparinized capillaries tubes.
32. Sealing wax or cristoseal.
33. Spirit lamp.
34. Microhematocrit reader scale.
35. Westergren's pipette.
36. Westergren's stand.
37. Trisodium citrate pentahydrate.
38. Wintrobe tubes.
39. Wintrobe tube stand.

Fig. 1 Experimental fishes being maintained in recirculating aquaculture systems (RAS)

4 Methods

4.1 Methods of Blood Sampling from Fish

4.1.1 Preparations

1. Fish are best maintained in recirculating aquaculture systems (RAS) with water exchange and biofiltration to reduce ammonia toxicity (Fig. 1).
2. Food should be withheld 24 h before blood sampling to avoid regurgitation and/or defecation during the procedure.
3. Hematology and clinical chemistry parameters from the collected blood should be obtained from living fish. Tissue fluids contamination should be avoided (*see* **Note 1**).
4. Equal to or less than 0.5–1.0% of the fish's body weight should be removed from a fish that will be recovered from blood collection.
5. Fish >200 g recover successfully after blood collection (dependent on the health status of the fish prior to blood collection).
6. Fish smaller than 200 g may have to be sacrificed to ensure an adequate blood volume is collected [13].
7. Sedate fish in bucket containing buffered 50 mg/L MS-222 (or similar anesthetic agent).
8. Remove from bucket once righting reflex is lost and place fish on tray or in a V-shaped holder.
9. Collect appropriate amount of blood (*see* **Note 2**) from different methods, which are described below.

Fig. 2 Numbering of fish prior to examination

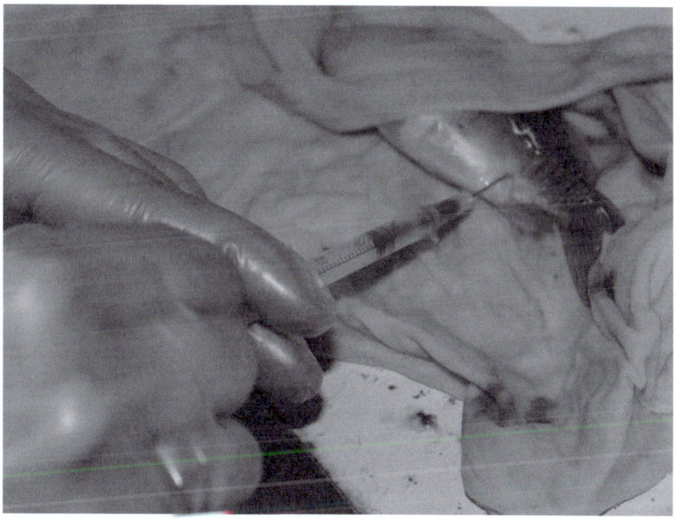

Fig. 3 Locating the caudal vein for blood sampling

4.1.2 Sampling

Blood may be sampled following five separate procedures: caudal vein, cardiac puncture, caudal severance, dorsal aorta, and kidney (*see* **Note 3**). Out of these, the caudal vein is the preferred method for freshwater fishes.

Caudal Vein

At the time of examination, fish is to be taken out from the aquaria, numbered, and placed carefully for blood extraction (Fig. 2). The caudal vein of the fish is located (Fig. 3) and blood is withdrawn in a heparinized syringe for examination (Fig. 4). Large fish are anesthetized with 0.2 mg/mL of tricaine methane sulfonate (MS-222, Sigma) (Poland) or CIFECALM (2 drops in 1 L of water) and

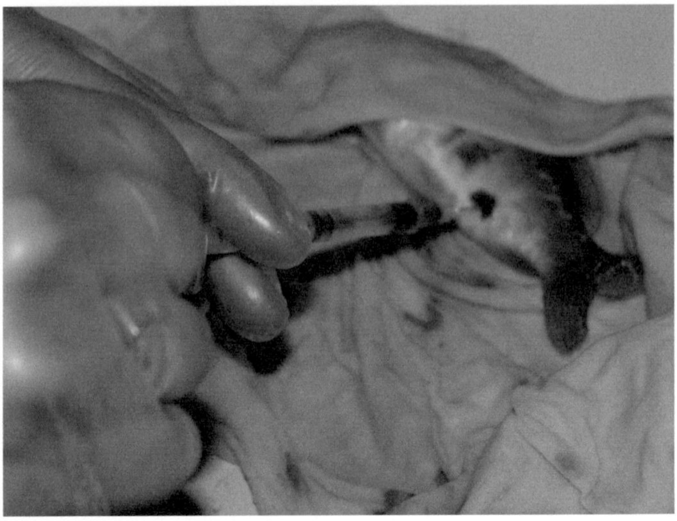

Fig. 4 Extracting blood from caudal vein

whole blood collected from the midline just posterior to the anal fin from caudal vein using 5-mL heparinized syringes. Insert the needle perpendicular into the musculature toward the ventral surface of the fish until the spine is reached or blood enters the syringe. If contact with the spine is made, withdraw the needle slightly. The vein is ventral to the overlying spine. This blood vessel can also be sampled laterally [14]. In either case, inserting the needle at a 45° angle is most effective for blood collection (*see* **Note 4**). The tested fish are reverted back into the aquaria.

Heart (Cardiac Puncture)

Blood is collected from the heart ventricle by gentle cardiac puncture or by using a thin Pasteur pipette introduced through the ventral body wall or below the pectoral fins [15]. The fish is sacrificed by severing the vertebral column just behind the skull. Insert the needle perpendicular into the musculature toward the ventral surface of the fish in the center of an imaginary line between the anterior most part of the base of the pectoral fins (*see* **Note 5**). A heparinized pipette or one containing a droplet of heparin saline solution is used occasionally for large fish to avoid clotting. For large fish, use of a Pasteur pipette to collect blood is not convenient, and when the fish are autopsied, the blood is also taken directly from the heart after the body cavity opening. Heparin is used as an anticoagulant.

The collected blood is examined by hanging drop preparations and microhematocrit for the presence of parasites.

Caudal Severance

Euthanize the fish that are too small to bleed with a syringe and needle. Caudal peduncle is dried. Posterior of the tail till the anal fin is completely severed. For good results, the fish should be seized with the head above the region of the tail and the microhematocrit tubule touched to the drop of blood pooling at the cut area (*see* **Note 6**). Fish is not squeezed to avoid sample contamination. Discard the first few drops and collect the rest in one or more microhematocrit tubules [8]. Fish is returned to a separate container after sample collection (*see* **Note 7**).

Dorsal Aorta

Insert needle on a 30–40° angle into the dorsal midline in the roof of the mouth at about the third to fourth gill arch. Depending upon size and species of fish insertion between the first and second arch may be more suitable [15]. After sample collection, recovering fish tend to bleed from the mouth. This site may be used for indwelling catheterization.

Kidney

The autopsied fish are used for this purpose. The kidney is extracted from the fish and minced tissue of kidney suspended in saline is used for parasite detection "kidney stroke" [16].

4.2 Examination of Blood

Blood parasites are best identified by microscopy as it still remains the dependable laboratory diagnosis for blood and tissue parasites. Diagnosis maybe hampered by sparseness of parasites on the slide and the differentiation between similar appearing organisms. Examination of thick and thin stained blood smears is employed for the detection and identification of species [17]. Serological assays are available in some cases but unfortunately, none of these can be solely used for diagnosis. Therefore, routine examination of blood still remains to be the popular method for hematological assays and blood parasite examinations.

4.2.1 Isolation and Processing of Blood Parasites

The blood can be examined either by hanging drop preparations to observe live parasites or by examining permanent stained smears.

Hanging Drop Preparation

This technique is used for first hand detection of trypanosomes in the blood. The technique requires a coverslip, a microscope, and a special slide having a central concavity. The blood sample suspected of containing the trypanosomes is diluted with a sterile isotonic solution. A drop of this fluid mixture is placed on a glass coverslip (Fig. 5), which is then inverted carefully and placed over the slide so that the drop hangs from the coverslip into the concavity in the slide [18]. The motile and delicate structure and its characteristic movement is viewed through the microscope for diagnostic purposes. A positive infection with hemoflagellates can be confirmed by the vigorous movement of the flagellates in fresh blood mounts, observed under the compound microscope at 40× magnification. The live flagellates can be identified not only by their constant motion but also by the undulating movement of the body continuously. The periodic contractions and elongations of the body changing the orientation of the anterior–posterior axis of the flagellates also aids in diagnosis.

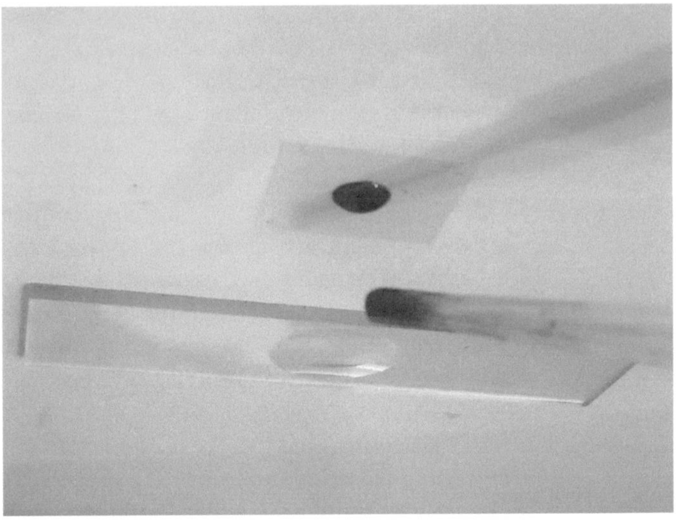

Fig. 5 Hanging drop preparation for examining live parasites in cavity slide

4.2.2 Concentration of Parasites

Clotting Method

When the infections are of extremely low intensity, the blood is allowed to clot in a centrifuge tube placed overnight in a refrigerator. The following day, it is spun at 13,000 × *g* for 10 min and the serum is discarded. Flagellates are revealed in the sediment even if parasitemia is extremely low. Conventional thin films are made of the pellet which contains trypanosomes and RBC. This is a modification using just a small clot of blood from which the flagellates escape into the serum, without later centrifugation [19]. However, this method is not a sure shot method of confirmation as subpatent chronic infection can not be confirmed by this method.

Microcentrifugation

Specially, the low-intensity infections of flagellates are also confirmed by the hematocrit centrifuge technique [20] by using microcentrifuge (REMI RM 12C) having maximum rpm of 16,000. A heparinized capillary tube is filled with blood (about 0.06–0.7 mL) and one end is sealed. The sealed tube is centrifuged in a hematocrit centrifuge for approximately 4 min at 12,000 rpm. The lid of the centrifuge is not completely closed in order to prevent heat buildup. The microcentrifuge is operated under appropriate safety precautions. After centrifugation, the parasites collect in the buffy layer (Fig. 6), the tube is carefully placed horizontally in a drop of immersion oil on a microscope slide and examined under the compound microscope at 1000× magnification. Infections are confirmed by detecting the parasites writhing at the junction of the buffy layer (the layer of packed blood cells) and the clear plasma. The tube is cut immediately above this layer, the suspension layered on slides, and stained with any of the techniques given below.

Fig. 6 Microcentrifuged blood showing buffy layer

The two-step centrifugation method: low-speed spinning removes most of the blood corpuscles (and some flagellates), while the following high-speed (2000 × *g*) centrifugation of the supernatant sediments, the flagellates. Centrifugation of heparinized blood in hematocrit tubes, which after spinning are broken at the erythrocyte–leukocyte interface and the buffy layer of white cells examined for live trypanosomes [21].

4.3 Blood Film Preparation

4.3.1 Microscope Slides

Clear and colorless slides which are corrosion–resistant, flat, free from distortions and ripples are used for preparing the blood films (*see* **Note 8**).

New slides are placed in potassium dichromate cleaning solution (20 g $Cr_2K_2O_7$ in 100 mL water with 900 mL concentrated H_2SO_4) for 48 h followed by a thorough rinse in running tap water. Slides are stored in 95% methanol and carefully wiped clean and dry before use. Used slides are soaked in a detergent at 60 °C for 15–20 min, rinsed in hot water before drying. Other cleaning agents used are chromic acid or 70% alcohol + 30% ether. The microscope lens and slides after being used with oil immersion are also cleaned in the latter solution.

Used slides (when used) are cleaned by soaking in a detergent at 60 °C for 15–20 min and rinsed in hot water before drying.

Subsequent to collection/concentration of blood, the smears are made and processed. The "Push" method is conveniently used for preparing the blood films and it is the preferred method as sufficiently large area is available for microscopic examination: this area shows all cells barely touching or separated from each other (monolayer part). The blood volume used is sufficient to draw a film 2.5–4.0 cm in length. The blood drop is placed at a distance of

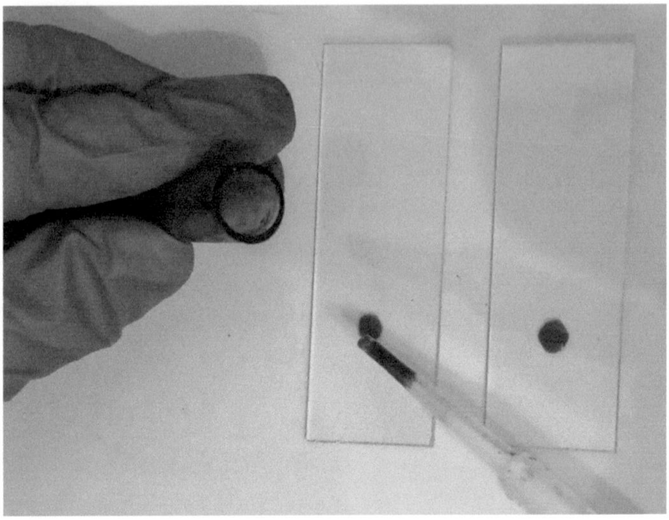

Fig. 7 Placing a drop of blood on a slide

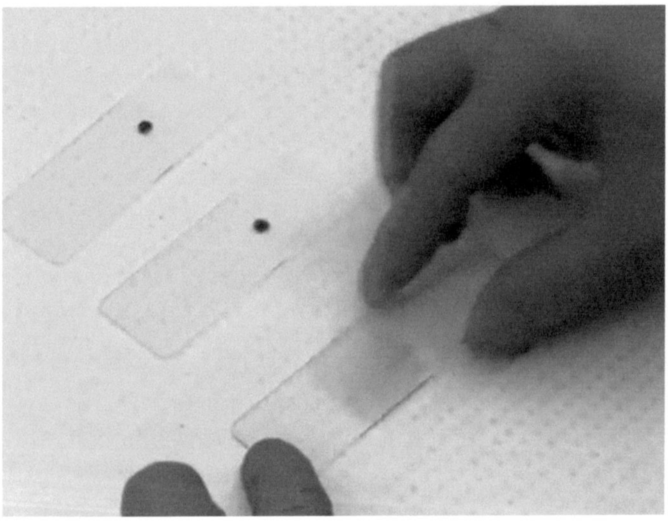

Fig. 8 Spreading of a blood film

1 cm from the slide end (Fig. 7). In cases of small volume of blood, the film is prepared by the cover glass method (*see* **Note 9**). Care is taken to use clean spreaders having smooth ends in order to ensure even thickness for the entire width of the blood film. After placing the blood drop on the slide, the drop is allowed to spread quickly along the spreader's edge. The spreader is now moved forward immediately at a steady rate, at a fairly fast speed and at 45° angle until all the blood is consumed (Fig. 8) into the film (*see* **Note 10**). The films are immediately air-dried [22].

Fixation and Staining

The air-dried films are immediately fixed in methanol (10 min) in order to avoid gray/blue background effects. Slides are stained by immersion in reagent filled Coplin jars. The blood films are stained by Romanowsky dyes and the following methods are employed:

Giemsa–Leishman Staining

The films are stained in Giemsa–Leishman + buffer (pH 6.5) in the ratio of 1:7 for 40 min [23].

JSB Staining

1. The methanol fixed slide is dipped in JSB II for 5 s, washed in buffer 10–20 times, dipped in JSB I for 30–40 s, washed in buffer, dried, and examined [24].
2. The positive slides are mounted in xylene-DPX having low viscosity so that there are no air bubbles and covered with cover glass to ensure safety of the slides. The slides are properly labeled and viewed under oil immersion in autophotomator microscope with digital photographic unit and photographed.

Hemacolor® (Merck) Rapid Blood Smear Staining Technique

Hemacolor® is a staining set for rapid manual staining of blood smears. The staining pattern of blood smear corresponds to classical staining patterns, cell elements and structures being stained distinctly. Morphological alterations can thus be easily recognized. The air-dried blood smears are fixed and stained by immersion into solutions 1–3 for 30 s [25].

1. Each solution is transferred into a separate cuvette with a lid or into a staining trough.
2. Blood smears are prepared and air dried.
3. The blood smear is immersed five times for 1 s each time into solution 1. The solution is allowed to drip off.
4. The blood smear is immersed three times for 1 s each time into solution 2.
5. The blood smear is immersed six times for 1 s each time into solution 3
6. The solution is allowed to drip off.
7. The smear is rinsed with buffer solution (pH 7.2). The buffer solution is prepared by dissolving 1 buffer tablet in 1 L of freshly prepared distilled water.
8. The slides are dried and examined under the microscope.

Wright Stain

1. The slides are fixed for 30 s in absolute methanol.
2. Methanol is removed by tilting the slide.
3. The slides are stained for 2 min on a horizontally positioned slide.
4. The aliquot of the buffer solution is added without any of the stain running off the slide. The buffer and stain is mixed gently

without touching the surface of the blood film on the slide. A metallic sheen appears on the surface of the staining solution mixture [26].

5. The slide is allowed to stand for 3 min.
6. It is rinsed with (distilled) water for 30 s.
7. It is dried in tilted position.
8. The positive slides are mounted in DPX.

Giemsa Azur Eosin Methylene Blue

Azur and eosin are acidic dyes which variably stain the basic components of the cells like the cytoplasm and granules, and methylene blue acts as the basic dye, which stains the acidic components, especially the nucleus of the cell.

1. This stain is either used readymade from MERCK or prepared as mentioned in Subheading 3.2.
2. Stain the fixed air-dried film, with diluted Giemsa Azur Eosin Methylene Blue (1:20, vol/vol) for 20 min.
3. For a 1:20 dilution, add 2 mL of stock stain to 40 mL of buffered water (pH 6.5) in a Coplin jar.
4. Wash by briefly dipping the slide in and out of a Coplin jar of buffered water (one or two dips). Excessive washing will decolorize the film.
5. Let air-dry in a vertical position.
6. Observe under the microscope first at 40x and then using oil immersion lens.

May–Grunwald–Giemsa Stain

1. The slides are fixed for 30 s in absolute methanol.
2. Methanol is removed by tilting the slide or by simply removing from the fixing jar.
3. Staining solution I is applied freshly diluted with an equal part of buffer for 5 min on a horizontally positioned slide or in a jar.
4. The slide is transferred from jar without washing (or the stain is removed by holding slide vertically) into staining solution II that is freshly diluted with 9 parts buffer for 10–15 min.
5. The slide is transferred to a jar with buffer for 1 rinse after removing stain [27].
6. The slide is washed with ample water.
7. The slide is transferred to a jar containing water for 2–5 min.
8. It is dried in tilted position.
9. The slide is mounted in DPX with a cover glass.

Gomori Trichrome Staining

This technique demonstrates the nucleus and kinetoplast of parasitic flagellates clearly with good results as it is more easily obtained than with the use of Heidenhain's iron hematoxylin staining [28].

1. Smears are fixed in Schaudinn's fluid for 15 min.
2. Wash for 2 min in 70% ethanol.
3. Wash for 3 min in distilled water.
4. Stain for 5–8 min in Harris's hematoxylin.
5. Hematoxylin is dissolved in alcohol and the alum in the water, the two solutions are mixed, rapidly boiled, mercuric oxide is added, and the solution is cooled. This solution is stable for many years.
6. The slides are then differentiated in acid alcohol (1 mL conc. HCl in 99 mL 70% ethanol) until only the nuclei are stained.
7. The slides are stained for 10 min (range of 5–20 min) in the following solution:

   Dissolve 0.6 g phosphotungstic acid in 100 mL distilled water, then 0.6 Chromotrope 2R, 0.3 g Fast green FCF, and 1 mL of glacial acetic acid is added.
8. The slides are rinsed in a solution of 0.2 mL glacial acetic acid in 100 mL of distilled water for 1 min.
9. The slides are dehydrated through an alcohol series, cleared in xylene, and mounted using a permanent mounting medium (nonviscous Canada balsam or DPX). Xylene can be used as a diluent if required.
10. The permanent preparations made as above are mounted in clear, nonviscous DPX (xylene is used as a diluent) and sealed. They are properly labeled and stacked in slide boxes.

4.4 Identification of Blood Parasites

4.4.1 Trypanosoma

Obligate parasite, unicellular flagellate protozoa, with one flagella and an elongated spindle-shaped leaf-like body. It has conspicuous nucleus varying in shape, size and position according to the species of the parasite [29]. A large clump of DNA located at one end of the unusually long mitochondrion represents the body is transversed with an undulating membrane having variable number of folds [30].

4.4.2 Trypanoplasma

The parasite has a cytostome and two flagella that arise from a flagellar pocket, an anterior and a posterior flagellum contractile (pulsatile) vacuoles at the base of flagellar pocket [31]. These biflagellated blood stages may be slender reaching a length of 15 μm or appear polymorphous-stumpy with diameters of up to 20 μm. The kinetoplast and nucleus are opposed to each other and the undulating membrane is not well marked. Both *Trypanosoma* and *Trypanoplasma* exhibit polymorphic stages [32].

4.4.3 *Haemogregarina*

Unicellular, elongate to fusiform oval organisms found in the red blood cells. Although the size varies, they are larger than the cell's nucleus [4]. The organism stains a basophilic color and has a surrounding clear zone. The mature gametocyte is the most commonly found stage: it is exoerythrocytic, but carries the host erythrocyte nucleus attached to its external surface near one end. The gametocyte has a central nucleus and 2–16 subterminal eosinophilic granules, but no polar cap [33].

4.4.4 *Lankesterella*

Comprises a type of plastid called an apicoplast, and an apical complex structure. The organelle is an adaptation that the apicomplexan applies in penetration of a host cell [5]. Merogony and sporogony occur in vascular endothelial cells in the visceral organs of the vertebrate hosts. Mature sporozoites are released into the bloodstream and invade erythrocytes.

4.4.5 *Hepatozoon*

The organism forms large (up to 11 μm) gametocytes, which are oval to elliptical, clear to pale blue structures within the cytoplasm of neutrophils, monocytes, or both. The sporozoites migrate to the liver of the vertebrate, where they undergo multiple fission (asexual reproduction) to produce merozoites. In the blood, the meronts released are visible where they form gametocytes, the final stage of development. The gamonts are large, conspicuous organisms which occupy a significant portion of the erythrocyte, and are easily visible in simple blood films [34].

4.4.6 *Plasmodium*

Intra erythrocytic stages consist of small rounded trophozoites (ring forms) measuring 1–2 μm in diameter, amorphous multinucleate schizonts measuring up to 7–8 μm in length, and micro- (♂) and macro- (♀) gametocytes ranging in length from 7 to 14 μm. The morphological characteristics (size, shape, and appearance) of the blood stages are characteristic for each *Plasmodium* spp. Microgametocytes have a larger more diffuse nucleus (ready for gamete production), while macrogametocytes have darker-staining cytoplasm (plentiful ribosomes for protein synthesis) [35].

4.4.7 *Haemoproteus*

Gametocytes are mainly present within erythrocytes. Organisms may appear similar to *Plasmodium*, but the pigment within the intraerythrocytic gametocytes is more dispersed and schizonts are not seen in the peripheral blood smears. These pigment granules (hemozoin) are derived from the over one half of the erythrocyte cytoplasm [36]. The parasite may cause slight enlargement of the infected host cells and displacement of the red blood cell nucleus to one side [37].

4.4.8 *Dactylosoma*

Schizonts in this genus produce 6–16 merozoites by simultaneous exogenous budding. These typically are found in a fan shaped arrangement. This occurs within the erythrocytes. Mature schizonts contain eight nuclei but undergo division only to two to four daughter cells. During cytoplasmic cleavage, schizonts assume triad, rosette, or cruciform shapes. Merozoites are finally produced through a series of binary fissions of these daughter cells which may also be involved in additional nuclear divisions [38].

4.4.9 *Babesiosoma*

Triangular or slightly rounded in shape. One end broad (occupied by the band shaped nucleus) and the other tapers bluntly. The cytoplasm is vacuolated and without pigment [39]. The majority of gamonts are much broader with two large, distinct vacuoles especially near the ends of parasite body, which are rather rounded.

4.4.10 *Leucocytozoon*

The nuclei of the schizonts is enclosed in a trilaminar membrane with peripherally arranged chromatin, contains numerous cytomers and multiple ribosomes [40].

4.4.11 *Myxobolus*

Triactinomyxon spores are made of a single style that is about 150 μm long and three processes or "tails," each about 200 μm long, three polar capsules, each of which contains a coiled polar filament between 170 and 180 μm long. A sporoplasm packet at the end of the style contains 64 germ cells surrounded by a cellular envelope. There are also three polar capsules, each of which contains a coiled polar filament between 170 and 180 μm long. Polar filaments in both this stage and in the myxospore stage rapidly shoot into the body of the host, creating an opening through which the sporoplasm can enter [41].

4.4.12 *Loma*

Parasite stages arranged in a random, unstratified manner in the xenoma; merogony by multiple fission; sporogonic stages isolated within a sporophorous vesicle containing several sporoblasts and polysporoblastic sporogony [42]. The spores appear to be isolated in a single vesicle, and the intervention of tubules that resolve in partitions of the initial sporophorous vesicle.

4.4.13 *Ichthyosporidium*

Radiating hyphae usually with a single dichotomy, growing from one of three types of central body. The large xenomas are filled with developmental stages and pleomorphic spores [43]. Wet mount preparations show two general spore types: microspore with mean length of 6.2 μm; and less numerous macrospores with mean length of 8.5 μm and mean width of 5.5 μm.

4.4.14 *Microsporidia*

Unicellular, produce highly resistant spore, coiled polar tube, tubule or filament, layered polaroplast, a posterior vacuole (thought to function as Golgi), and a protective exospore made of proteins and chitin and is about 1–4 μm in size [44, 45].

4.4.15 *Pseudoloma*

Intracellular, fungus like appearance having distinct axonal swellings mainly found in microsporidium of zebrafish [46]. Xenomas are composed of several aggregates of up to 16 uninucleate spores segregated within sporophorous vesicles. Spores are oval to pyriform and measure approximately 5.4 × 2.7 μm with 13–16 coils of the polar filament [47].

4.4.16 *Spraguea*

Dimorphic, having two spore types: a large spore (4.0 × 1.25 μm) containing a diplokaryon and three to four polar tube coils and a smaller uninucleate spore (3.5 × 1.5 μm) with five to six polar tube coils [48, 49].

4.4.17 *Microgemma*

The parasite develops in liver but in the blood, polaroplast is bipartite and isofilar polar filament is coiled with 8–9 turns in a single or double row at the posterior end of the spore [50]. The nucleus is voluminous and central in position, measuring ~0.9 μm in diameter. A large posterior vacuole appears as pale area, occupying about a third of the spore length [51].

4.4.18 *Thelohania*

Monomorphic ovoid spores (5 × 4 μm) bound by a dense membranous exospore wall overlaying a thick lucent endospore wall. Mature spores are unikaryotic and contain an isofilar polar tube arranged in 17–19 coils in two layers [52].

4.5 Methods for Determination of Hematological Paramoters

4.5.1 Calculation of Differential Leucocyte Count (D.L.C)

Differential leukocyte counts belong to important characteristics of the health state of fish and in many cases, they are also helpful in evaluating the immune system. It is the most important investigation in hematology which provides information about cells, their number, shape, size, and variations in morphology to diagnose different types of anemias and other hematological disorders [53]. Differential leucocyte count (DLC) is important in diagnosis of leukocytosis and leucopenia and nonhematologic diseases. It is also used for the detection of certain parasites.

1. **Preparation of the thin blood film** (follow as mentioned in Subheading 4.2.2.1).
2. **Fixation and staining** (follow as mentioned in Subheading 4.2.2.2).
3. **Differentiation and counting**.

   The cell should be counted using high power or oil immersion lens in a strip running the whole length of the film. The lateral edges of the film are avoided and inspected from head to the tail. If less than 100 cells are encountered in a single narrow strip, one or more additional strips should be examined until at least 100 cells have been counted. Different types of leukocytes like neutrophils, eosinophils, basophils, monocytes, and lymphocytes are present in peripheral blood film. Mainly

Table 1
Hematological parameters of some common freshwater fishes of India

Hematological parameters	*Channa punctatus* [63]	*Heteropneustes fossilis* [64]	*Clarias batrachus* [64]	*Labeo rohita* [65]	*Catla catla* [66]	*Cirrhinus mrigala* [66]
RBC ($\times 10^6$ mm^3)	3.42	3.1	2.97	1.01	1.18	1.32
WBC ($\times 10^3$ mm^3)	4.3	9.84	7.63	6.64	18.5	11.4
Hb (%)	10.84	11.6	9.5	2.66	9.5	8.6
PCV (%)	36.14	31.48	30.2	14.0	21.0	25.0
ESR (mm/h)	5.50	6.0	5.21	5.0	4.0	4.5
MCV(cμ)	105.6	110.46	109.64	137.2	68.0	128.0
MCH (pg)	21.09	38.05	33.25	26.3	41.0	40.0
MCHC (%)	29.99	31.95	34.94	18.9	47.0	34.0

polymorphonuclear neutrophils and monocytes predominate at the margins and the tail and lymphocytes in the middle of the film. This separation depends upon differences in stickiness, size, and specific gravity among the different classes of cells (Table 1).

Leukocytes are differentiated by the morphological characteristics [54, 55] of the nucleus and are characterized as follows:

Lymphocytes

Majority of lymphocytes in the blood are small but large lymphocytes are also found. The cytoplasm of a lymphocyte is scanty, with few unevenly distributed granules around them. The nucleus is generally round, oval, or slightly indented, and the chromatin (a network of fibers within the nucleus) is lumpy and condensed at the periphery. It is believed that lymphocytes produce antibodies and destroy the toxic products of protein metabolism.

Monocytes

The monocyte is the largest of the normal white blood cells. The nucleus is lobulated, deeply indented or horseshoe-shaped, and has a relatively fine chromatin structure. The monocyte destroys bacteria, foreign particles, and protozoa.

Neutrophils

Neutrophils account for the largest percentage of leukocytes and function by ingesting invading bacteria. Its cytoplasm has numerous fine, barely visible granules and nucleus. The nucleus may be oval, horseshoe, or "S" shaped, or segmented (lobulated).

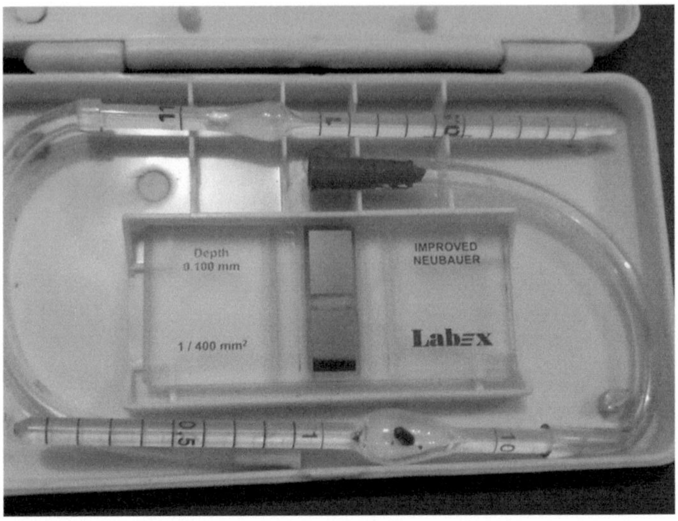

Fig. 9 Neubauer Hemocytometer for counting RBC and WBC

Eosinophil

The cytoplasm of an eosinophil contains numerous large ovoid specific granules, each containing elongated crystalloid regular discoidal shape, bilobed nucleus, placed eccentrically in the periphery. Eosinophils aid in detoxification. They also break down and remove protein material.

Basophil

Scattered large granules that are darker than the nucleus, characterize the cell as basophil granules may overlay the nucleus as well as the cytoplasm. Basophilic cells prevent the blood from clotting in inflamed tissue.

4.5.2 Calculation of Red Blood Cell Count (Erythrocytes)

1. This is the method of counting total number of erythrocytes in a calibrated chamber by dilution of blood to 1 in 200 with a diluent which is isotonic to blood.
2. Observation procedures and counting of red blood cell counts is done by thinning the blood with Hayem's solution in a maximum scale mixing pipette 101 by Neubauer's hemocytometer (Fig. 9).
3. There is a red grain inside this pipette that functions as a stirrer. The blood is extracted with a mixing pipette up to a scale of 0.5, then with the same pipette, Hayem's solution is sucked to scale 101.
4. The pipette is shaken for 3–5 min so that the blood is evenly mixed (*see* **Note 11**).
5. The first drop is removed and the next drop is dropped into the hemocytometer and covered with a cover glass (*see* **Note 12**).
6. Charge the chamber till it is completely charged by the capillary action.

7. Allow the cells to settle in the chamber for 2–5 min.
8. Count RBCs using high power objective. Calculations are carried out in 5 small boxes of the central chamber, namely, the upper left corner, upper right corner, lower left corner, lower right corner, and middle of Neubauer's chamber, and the counts are added—*N* [56].

$$\text{RBCs count} = \frac{N}{1/5 \times 1/10 \times 1/200} = \text{counted cells} \times 10,000$$

where,

N = Number of RBCs in 5 small boxes.

$1/200$ = Dilution of blood.

$1/10$ mm = Depth of chamber.

$1/5\text{mm}^2$ = Area counted ($5 \times 1/5 \times 1/5\ \text{mm}^2$).

9. Therefore, the calculated number of red blood cells are converted by the formula:
   Red Blood Cells count = Σ counted red blood cells × 10^4 cells/mm^3 [57].

4.5.3 Calculation of White Blood Cell (Leukocytes)

1. Blood is diluted with WBC fluid that hemolysis the red cells.
2. A blood sample is sucked with a pipette containing a white stirrer scale up to a scale of 0.5, then Turk's solution is added to a scale of 1 L.
3. Stirring is done in a pipette by swinging the pipette in the form of figure eight in air for 3–5 min until the blood is evenly mixed (*see* **Note 13**).
4. Remove the first drop of the blood solution in the pipette, drop the blood sample on the Neubauer's chamber [56].
5. Wait for 2–3 min for cells to settle.
6. Charge the chamber by placing the tip of the pipette just under the coverslip and fluid flows under it by capillary action till the counting chamber just filled.
7. Examine the ruled area under the microscope using low power objective.
8. Count white cells in 4 corner large squares.

$$\text{WBCs count} = \frac{N \times 20}{4^* \times 0.1} = N \times 50$$

where,

N = Number of WBCs in 4 large squares.

0.1 mm = Depth of chamber.

Dilution = 1 in 20.

4^* = Corner squares are counted.

9. Therefore, the total number of WBCs are counted with the following formula:

   The number of calculated leukocyte cells × 50 cells/mm^3 [57].

4.5.4 Hemoglobin (Hb) Measurement

Sahli's Acid Hematin Method

1. This is the most popular color matching method for Hb estimation.
2. The principle of this method is to convert hemoglobin in blood to hematinic acid by hydrochloric acid. Brown color of acid hematin is matched against the brown color of comparator.
3. The blood is sucked using a Sahli pipette to a scale of 20 cu mm, the tip of the used pipette is cleaned with tissue paper.
4. The blood is then transferred into the hemoglobin tube containing HCl 0.1 N to scale 2, then left for 3–5 min so that hemoglobin reacts with HCl to form hematinic acid [58]. The blood is then stirred with distilled water and added until the color becomes the same as the standard color (*see* **Note 14**).
5. Scale reading is done by looking at the surface height of the solution matched with a G% lane scale, which shows the amount of hemoglobin in grams per 100 mL of blood (*see* **Note 15**).

Cyanmethemoglobin Method

1. This method is based on the principle that Hb is oxidized to methemoglobin by potassium ferricyanide. MetHb in turn is converted to stable cyanmethemoglobin by Potassium cyanide. Color of this solution is compared against a standard of known Hb value calorimetrically.
2. Take 5 mL of Drabkin's solution in a test tube.
3. Add 0.02 mL of blood using a Hb pipette (as in Sahli's method). Rinse the pipette twice by drawing in Drabkin's solution. Mix blood with Drabkin's solution thoroughly (*see* **Note 16**).
4. Wait for 10 min.
5. Switch on the colorimeter with a green filter/540 nm and wait for 5 min.
6. Adjust optical density (OD) at 0 using Drabkin's solution as blank.
7. Pour test solution into the cuvette and note down OD of the test sample [59].
8. OD of the standard cyanmethemoglobin solution is taken against the blank.

$$\text{Hb} = \frac{\text{OD of the test sample} \times \text{conc. of std.}^{*} \times 250^{**}}{\text{OD of std.} \times 1000^{***}}\ \text{g\%}$$

* Conc. of standard is given for each standard (e.g., 50 mg/100 mL).

** Since 0.02 mL of blood has been added to 5 mL Drabkin's, dilution is 1 in 250.

*** Divide by 1000 to get the value in g%.

Hematocrit (PCV or Packed Cell Volume)

Microhematocrit Method

1. Hematocrit is the % volume of red cells (when packed) in a given sample of blood. The blood is sucked using a heparin-coated microhematocrit tube with a capillary system (*see* **Note 17**).
2. The function of heparin is to prevent blood clots in the tube.
3. After the blood reaches ¾ part of the tube, then one end of the tube is blocked with cristoseal or sealing wax (*see* **Note 18**).
4. Place the sealed tube into one of the numbered grooves of a microhematocrit centrifuge.
5. The capillary tube which fills with blood is then rotated using Microhematocrit centrifuge at 10,000 rpm for 5 min [60].
6. After the centrifuge has stopped, remove the specimen tube. The red blood cells have been packed into the bottom of the tube. The clear liquid on top of the cells is plasma.
7. Measurements are done by comparing blood body volume to the volume of all blood using a microhematocrit reader scale (*see* **Note 19**).

Wintrobe Method

1. Anticoagulated whole blood is centrifuged and the volume of red cell mass denotes the hematocrit, expressed as percentage [61].
2. Collect 2 mL blood in an EDTA vial.
3. Take Wintrobe's tube. Shake the blood sample so that plasma and cells are mixed properly.
4. Take blood in Pasteur's pipette from the vial and place the end of nozzle to the bottom of the lumen of Wintrobe tube (*see* **Note 20**).
5. Press the teat of the Pasteur's pipette very slowly simultaneously start withdrawing it from the tube, so that fills bottom and gradually reaches the top at 100 marking.
6. Centrifuge the tube at 3500 rpm for 30 min.
7. Take out Wintrobe tube and read the column of packed cells on PCV scale (e.g., 0 from bottom upward).

Estimation of Erythrocyte Sedimentation Rate (ESR)

When anticoagulated blood is allowed to stand vertically, sedimentation of erythrocytes occurs leaving clear plasma as the supernatant. The rate of sedimentation estimated under standard conditions is known as ESR. Sedimentation take place in three stages: formation, sinking, and packing of rouleaux.

Westergren's Method

1. Collect 2 mL of the blood in 0.5 mL of 3.8% Trisodium citrate pentahydrate solution (*see* **Note 21**).
2. Mix the two and withdraw blood by rubber teat attached on top of Westergren's pipette up to 0 mark.
3. Place the pipette on rubber cork of Westergren's stand and make it vertical using the adjustable levelling screws and detach the rubber teat.
4. Measure the height of column after 1 h. This gives the ESR in mm in first hour [62].

Wintrobe Tube Method

1. Collect 2 mL of blood in an EDTA vial.
2. Fill Pasteur pipette having a long stem.
3. Tip of the pipette is taken to the bottom of the Wintrobe tube and blood is slowly filled into the tube.
4. The pipette is slowly withdrawn and the blood is filled up to top mark of 0/100 [61].
5. Keep the tube in vertical position and leave undisturbed in a Wintrobe tube stand. This allows the sedimentation of erythrocytes (*see* **Note 22**).
6. After 1 h, the level of fall of red cells from top mark of 0 is noted as mm/hour.

The Derived Hematological Indices of MCV, MCH, and MCHC Are Calculated by Using Standard Formulae as Follows

1. **MCV** (mean corpuscular volume) is calculated in cubic microns (cμ) = vol. packed red cells, mL per 1000 mL/red cell count, millions per cu mm × 10;
2. **MCH** (mean corpuscular hemoglobin) is calculated in picograms (pg) = hemoglobin, gm per 1000 mL/red cell count, millions per cu mm;
3. **MCHC** (mean corpuscular hemoglobin concentration) is calculated in percentage (%) = hemoglobin, gm per 100 mg mL × 100/vol. packed red cells, mL per 100 mL.

5 Notes

1. Always wear gloves while handling blood collection equipment and blood. This will help to reduce the risk of contamination [2].
2. Fish must be allowed adequate time to recover and regenerate blood volume if serial blood samples are to be collected [2].
3. Blood samples should be processed within 30–60 min after collection. The time between blood sampling and laboratory evaluation of the sample collected is critical. Cells can swell and

rupture, and some parameters may not be stable during the time between collections and processing [11].

4. Remove the needle carefully and whole blood is emptied instantly into the suitable microcentrifuge tube. Syringe with needle may lead to lysis of red bloods cells.
5. Resheath of needles is prohibited. Incinerate and discard them into containers immediately after use.
6. Clean, spacious, and contamination-free working environment is recommended.
7. Overfilling of the container should be avoided. Follow the recommendations of fill line on the container.
8. Ensure usage of scratch-free slides which are free of dust, lint, and fingerprints, and dry at the time of use.
9. Blood films should be prepared within a couple of hours of blood sample collection to avoid artifactual changes that will distort the morphology of blood cells [2, 11].
10. If film is made too thin or if a rough-edged spreader is used many of the leukocytes, perhaps even 50% of them accumulate at the edges and in the tail. Moreover, a gross qualitative irregularity in distribution is the rule. Polymorphonuclear neutrophils and monocytes predominate at the margins and the tail and lymphocytes in the middle of the film. This separation probably depends upon differences in stickiness, size, and specific gravity among the different classes of cells [2, 11, 12].
11. When blood is taken from an EDTA vial, the sample should be mixed thoroughly, otherwise high/low counts may be obtained if mainly sedimented cells/plasma is taken.
12. Whenever possible, both sides of the chamber should be filled and count carried on both sides and the mean of the two taken.
13. The sample should be mixed thoroughly, otherwise high/low counts may be obtained. Clots in the blood entangle cells giving fallacious results.
14. If blood in EDTA vial is not mixed properly before filling the pipette, low/high value may be obtained depending on whether mainly the plasma or the sedimented RBCs are taken in the pipette [11, 12].
15. After 10 min, color of acid hematin starts fading; lower Hb value is obtained if reading is taken thereafter.
16. Never pipette Drabkin's solution by mouth; always use a suction bulb/autodispenser.
17. In case of blood spill; paper towels can be used to soak up the blood, remaining blood is cleaned and the area is sprayed with a 70% alcohol solution and is allowed to sit for 10 min, repeat if necessary [12, 32].

18. The tube's sealed end should point outward from the center and should touch the rubber lining on the rim of centrifuge. The centrifuge should be balanced by placing specimen tubes on opposite sides of the moving head, the inside cover should be tightened with lock wrench, and the outside cover should be securely fastened [11, 12].
19. Microhematocrit reading should be taken quickly, as slanting of red cell column occurs if tubes are left in a horizontal position for more than 5 min. Take care not to include buffy coat while recording the PCV [2, 11].
20. There should no bubble in the blood column.
21. The ratio of anticoagulant blood should be 1:4.
22. The tube should not be inclined, it should be vertical. There should not be any clot in the blood sample.

    Hematology has some limitations as an indicator of disease in fish because:
23. Blood counts, other than total cell count, cannot be automated due to the nucleated erythrocytes and nucleated thrombocytes [61].
24. For many species only relatively small volumes are readily collected [61].
25. As fish blood is poikilothermic, blood cell enzymes operate over a larger temperature range. Thus removal of blood from the host does not result in the marked slowing of enzymatic activity seen with mammalian blood, even if the samples are held on ice (though this will reduce enzyme activity). Consequently, fish blood cells continue to be active, both enzymatically and osmotically, which may result in large changes during transport of mean erythrocyte volume and packed cell volume [9].

References

1. Simide RM, Richard S, Alvis NP et al (2016) Assessment of the accuracy of physiological blood indicators for the evaluation of stress, health status and welfare in Siberian sturgeon (*Acipenser baerii*) subject to chronic heat stress and dietary supplementation. Int Aquat Res 8:121–135
2. Gupta DK, Gupta N, Gupta A (2001) Infectivity and pathogenecity in *Clarias batrachus* parasitized by haemoflagellates. Appld Fish Aqua 1(1):75–77
3. Cook CA, Sikkel PC, Renoux LP et al (2015) Blood parasite biodiversity of reef-associated fishes of the eastern Caribbean. Mar Ecol Prog Ser 533:1–13
4. Rabie SAH, Hussein NA (2014) A description of *Haemogregarina* species naturally infecting white-spotted gecko (*Tarentola annularis*) in Qena. J Egypt Soc Parasitol 44(2):351–358
5. Paperna I, Ogara W (1996) Description and ultrastructure of *Lankesterella* species infecting frogs in Kenya. Parasite 4:341–349
6. Meraj M, Nizam M, Wani SA et al (2017) Alteration in hematology of *Cyprinus carpio* under the stress of pollution of water bodies of Kashmir valley. Int J Fish Aquat 5 (5):176–179
7. Kefas M, Abubakar KA, Jaafaru A (2015) Haematological indices of tilapia (*Oreochromis*

niloticus) from Lake Geriyo, Yola, Adamawa state, Nigeria. Int J Fish Aquat. 3(1):09–14
8. Clark TD, Donaldson MR, Drenner SM et al (2011) The efficacy of field techniques for obtaining and storing blood samples from fishes. J Fish Biol 79(5):1322–1333. https://doi.org/10.1111/j.1095-8649.2011.03118.x
9. Francesco F (2019) Fish hematology analysis as an important tool of aquaculture. Rev Aquacult 500:237–242
10. Eyiwumi FA, Augustine O, Ovie AK (2018) The hematological parameters of catfish (*Clarias gariepinus*) fed fish feeds with replaced premix using Moringa leaf meal (MLM). Madridge J Aquac Res Dev 2(1):35–39. https://doi.org/10.18689/mjard-1000107
11. Nigar S, Gupta N, Shalaby SI (2017) Haematological indices of *Channa punctatus* as a bio-indicator assessing physiological status. Bull NRC 41(2):49–61
12. Nigar S, Gupta N (2017) Heavy metals: implications associated to fish consumption. In: Pandey BN (ed) Recent advances in aquaculture. Narendra Publishing House, Delhi, pp 99–107
13. Blaxhall PC, Daisley KW (1973) Routine haematological methods for use with fish blood. J Fish Biol 5:771–781
14. Argungu LA, Siraj SS, Christianus A et al (2017) A simple and rapid method for blood collection from walking catfish, *Clarias batrachus* (Linneaus, 1758). Iran J Fish Sci 16 (3):935–944
15. Canada department of fisheries and ocean Animal user training Template (2004). https://www.ccac.ca/Documents/Education/DFO/4_Blood_Sampling_of_Finfish.pdf. Accessed Sept 2004
16. Snedecor GW, Cochran WG (1989) Statistical methods. Oxford and IBH Pub Co, Calcutta
17. Rosenblatt JE (2009) Laboratory diagnosis of infections due to blood and tissue parasites. Clin Infect Dis 49(7):1103–1108. https://doi.org/10.1086/605574
18. Pradhan P (2015) Hanging drop test/preparation. https://microbesinfo.com/2015/02/hanging-drop-testpreparation/. Accessed 15 Feb 2015
19. Strout RG (1962) A method for concentrating hemoflagellates. J Parasitol 48:100
20. Woo PTK (1969) The hematocrit centrifuge for the detection of trypanosomes in blood. Can J Zool 47:921–923
21. Cottrell BJ (1977) A trypanosome from the plaice *Pleuronectes platessa* L. J Fish Biol 11:35–47
22. Adewoyin AS, Nwogoh B (2014) Peripheral blood film—a review. Ann Ib Postgrad Med 12(2):71–79
23. Gajendra S, Jha B, Goel S et al (2015) Leishman and Giemsa stain: a new reliable staining technique for blood/bone marrow smears. Int J Lab Hematol 37(6):774–782
24. Manwell RD, Feiglson P (1948) A modified method of preparing the JSB stain. J Lab Clin Med 33(4):777–782
25. Herreros MG, Aparicio IM, Baron FJ et al (2006) Standardization of sample preparation, staining and sampling methods for automated sperm head morphometry analysis of boar spermatozoa. Int J Androl 29:553–563
26. Hoppe BR, Lassen ED (1978) Blood smears and the use of Wright's stain. Iowa State Univ Vet 40(3):113–116. https://lib.dr.iastate.edu/iowastate_veterinarian/vol40/iss3/10
27. Harashima S (1986) On the historical review of the Giemsa's, may-Grunwald-Giemsa and Wright's stain. J Jpn Soc Clin Cytol 25 (4):602–609
28. Gomori trichrome stain protocol (2006). https://neuromuscular.wustl.edu/pathol/histol/GT.pdf. Accessed 26 Dec 2006
29. Gupta DK, Gupta N (2010) A haemogram of *Clarias batrachus* parasitized by two species of haemoflagellates (*Trypanosoma batrachi* Qadri, 1962 and *Trypanoplasma haematalis* n. sp.) as an indicator of infectivity. Rev Fish Sci 18 (2):177–182
30. Gupta DK, Gupta N (1996) Potentialities of haemoflagellates in the degradation of blood parameters of *Mystus vittatus*. In: Bhatnagar SK (ed) Recent advances in biotechnology, Society for Plant Research, pp 122–127
31. Woo PTK (2003) *Cryptobia* (*Trypanoplasma*) *salmositica* and salmonid cryptobiosis. J Fish Dis 26:627–646
32. Gupta DK, Gupta N (1999) A new record of *Trypanoplasma* (Kinetoplastida: Cryptobiidae) inhabiting the blood of carp, *Catla catla* in India. Bull NRC 24:75–83
33. Hayes PM, Smit NJ (2019) Molecular insights into the identification and phylogenetics of the cosmopolitan marine fish blood parasite, *Haemogregarina bigemina* (Adeleorina: Haemogregarinidae). Int J Parasitol Parasites Wildl 8:216–220
34. Pereira GR, Soares P, Gomes MQ et al (2013) Are fish paratenic natural hosts of the caiman haemoparasite *Hepatozoon caimani*? Parasitol

Res 113(1):39–45. https://doi.org/10.1007/s00436-013-3623-9

35. Collins WE, Jeffery GM (2007) *Plasmodium malariae*: parasite and disease. Clin Microbiol Rev 20(4):579–592
36. Karadjian G, Puech MP, Duval L et al (2013) *Haemoproteus syrnii* in *Strix aluco* from France: morphology, stages of sporogony in a hippoboscid fly, molecular characterization and discussion on the identification of *Haemoproteus* species. Parasite 20:32. https://doi.org/10.1051/parasite/2013031
37. Gupta N, Gupta DK, Yadav P (1998) New records of some haemoparasites affecting *Puntius ticto* (Pisces: Cyprinidae) in India: observations on interaction and displacement of species. Res Rev Parasitol 58:103–108
38. Paperna L (1981) *Dactylosoma hannesi* n. sp. (Dactylosomatidae, Piroplasmia) found in the blood of grey mullets (Mugilidae) from South Africa. J Parasitol 28(4):486–491. https://doi.org/10.1111/j.1550-7408.1981.tb05325.x
39. Smit JN, Van JG, Davies AJ (2003) Observations on *Babesiosoma mariae* (Apicomplexa: Dactylosomatidae) from the Okavango delta, Botswana. Folia Parasitol 50:85–86
40. Lee HR, Koo BS, Jeon EK (2016) Pathology and molecular characterization of recent *Leucocytozoon caulleryi* cases in layer flocks. J Biomed Res 30(6):517–524
41. Benoit LFG, Sorel DN, Abraham F (2017) Three new species of *Myxobolus* (Myxosporea: Myxobolidae), parasites of *Barbus callipterus* Boulenger, 1907 in Cameroon. Asian J Biol Sci 10:110–112
42. Fomena A, Coste F, Bouix G (1992) *Loma camerounensis* sp. nov. (Protozoa:Microsporida) a parasite of *Oreochromis niloticus* Linnaeus, 1757 (Teleost:Cichlidae) in fish-rearing ponds in Melen, Yaoundé, Cameroon. Parasitol Res 78(3):201–208
43. Porter A, Vinall HF (2006) A protozoan parasite (*Ichthyosporidium* sp.) of the neon fish *Hyphessobrycon innesi*. J Zool 126(3):397–402. https://doi.org/10.1111/j.1096-3642.1956.tb00446.x
44. Ramanan P, Pritt BS (2014) Extraintestinal microsporidiosis. J Clin Microbiol 52(11):3839–3844
45. Garcia LS (2002) Laboratory identification of microsporidia. J Clin Microbiol 40:1892–1901
46. Ferguson JA, Watral V, Adam R et al (2007) Spores of two fish Microsporidia (*Pseudoloma neurophilia* and *Glugea anomala*) are highly resistant to chlorine. Dis Aquat Org 76:205–214
47. Spagnoli ST, Xue L, Murray KN (2015) *Pseudoloma neurophilia*: a retrospective and descriptive study of nervous system and muscle infections, with new implications for pathogenesis and behavioral phenotypes. Zebrafish 12:2. https://doi.org/10.1089/zeb.2014.1055
48. Casal G, Clemente CS, Matos P et al (2012) Redefining the genus *Spraguea* based on ultrastructural and phylogenetic data from *Spraguea gastrophysus* n. sp. (phylum Microsporidia), a parasite found in *Lophius gastrophysus* (Teleostei) from Brazil. Parasitol Res 111:79–88
49. Freeman MA, Yokoyama H, Osada A (2011) *Spraguea* (Microsporida: Spraguidae) infections in the nervous system of the Japanese anglerfish, *Lophius litulon* (Jordan), with comments on transmission routes and host pathology. J Fish Dis 34(6):445–452
50. Casal G, Matos E, Gracia P, Quraishy SA et al (2012) Ultrastructural and molecular studies of *Microgemma carolinus* n.sp. (Microsporidia), a parasite of the fish *Trachinotus carolinus* (Carangidae) in Southern Brazil. Parasitology 139(13):1720–1728
51. Mansour L, Prensier G, Jemma SB (2005) Description of a xenoma-inducing microsporidian, *Microgemma tincae* n. sp., parasite of the teleost fish *Symphodus tinca* from Tunisian coasts. Dis Aquat Org 65(3):217–226
52. Moodie EG, Jambre LF, Katz ME (2003) *Thelohania parastaci* sp. nov. (Microspora: Thelohaniidae), a parasite of the Australian freshwater crayfish, *Cherax destructor* (Decapoda: Parastacidae). Parasitol Res 91(2):151–165
53. Gupta N, Gupta DK (1990) Trypanosomes as etiological agents of anaemia in a freshwater siluroid, *Wallago attu*. In: Hirano R, Hanyu I (eds) Proc.II Asian fish. Forum, 1st edn. A.F.S, Philippines, pp 709–711
54. Tripathi A (2014) Cytological study on the leukocytes of selected fresh water fishes of India. Int J Fish Aquat Stud. 2(1):17–23
55. Shahi N, Mallik SK, Sarma D (2014) Leukocyte response and phagocytic activity in common carp, *Cyprinus carpio* experimentally infected with virulent *Aeromonas allosaccharophila*. J Ecophysiol Occup Hlth 14:66–70
56. Chaturvedi KU, Tejinder S (2002) Practical pathology. Arya Publications, New Delhi, pp 56–60
57. Jain NC (1986) Schalm's veterinary hematology, 4th edn. Lea & Febiger, Philadelphia, pp 20–45

58. Talib VH, Khurana SR (1999) A handbook of medical laboratory technology, 2nd edn. CBS Publisher and Distributer, New Delhi, pp 3–14
59. Balasubramaniam P, Malathi A (1992) Comparative study of hemoglobin estimated by Drabkin's and Sahli's methods. Postgrad Med J 38(1):8–9
60. Wennecke G (2004) Hematocrit- a review of different analytical methods, https://acutecaretesting.org/en/articles/hematocrit%2D%2Da-review-of-different-analytical-methods. Accessed Sept 2004
61. Wintrobe MM (1981) Clinical hematology, 6th edn. Lea & Febiger, Philadelphia, pp 410–439
62. Mechatronics RR (2015) A classic, gold standard: The Westergren method for ESR measurement. https://rrmechatronics.com/wp. Accessed Aug 2015
63. Malathi K, Kannathasan A, Rajendran K (2012) Comparative haematological studies on fresh water fishes *Channa punctatus* and *Channa striatus* (Bloch). Int J Pharm Chem Biol Sci 2(4):644–648
64. Acharya G, Mohanty PK (2014) Comparative haematological and serum biochemical analysis of catfishes *Clarias batrachus* (Linnaeus, 1758) and *Heteropneustes fossilis* (Bloch, 1794) with respect to sex. J Entomol Zool Stud 2 (6):191–197
65. Jasmin J, Rahman MM, Rahman MM (2018) Haematological changes in *Labeo rohita* (H.) due to exposure of pesticides, difenoconazole and thiamethoxam. Int J Contemp Res Rev 9 (1):20199–20205
66. Southamani C, Shanthi G, Deivasigamani M (2015) Hematological response in three Indian major carps in relation to supplementary feeding. Int J Fish Aquat Stud 3(2):287–294

Chapter 5

Techniques to Conduct Morphological and Molecular Investigations on Nematodes

Aasha Rana, Anshika Yadav, Aashaq Hussain Bhat, Ashok Kumar Chaubey, and Sandeep K. Malhotra

Abstract

The intriguing variety in the form and size of roundworms leading free-living as well as parasitic life has mesmerized scientists world over. One most common happening during developmental cycle of roundworms is moulting, and this characteristic is remarkably retained irrespective of the presence or absence of intermediate hosts (direct life cycle) in the life cycle of round worms in certain specified cases. The application of fixatives, in particular, therefore becomes a challenging exercise to suit the body form at different stages of development. However, looking at the severity of pathogenic influence of animal-parasitic and plant-parasitic nematodes necessitated appropriate fixation procedures of these worms infesting separate variety of hosts. Such procedures are enumerated in the text of this chapter with meticulous details on morphology based on scanning electron microscopic examination of certain worms, wherever required. Prominent techniques with molecular applications in taxonomy of roundworms to segregate arthropod-parasitic, freshwater, marine, and soil taxa have been outlined. The methods for effective barcode analyses of these round worms from terrestrial or aquatic vertebrate hosts have been discussed briefly.

Keywords Nematodes, Liver habitat, Molecular taxonomy, Plant-parasitic nematodes, Entomopathogenic, Zoonotic, DNA barcoding, Genomic DNA extraction

1 Introduction

Nematodes are multicellular tiny roundworms with worldwide distribution. The plant-parasitic nematodes (PPNs) are categorized as ectoparasitic (complete their life cycle in soil), endoparasitic (complete their life cycle inside the roots or in plant), and semi-endoparasitic (partial endoparasitic and partial free living or both). The beneficial ones are mainly insect-parasitic nematodes commonly called entomopathogenic nematodes (EPNs) and are commercially used as biopesticides to replace hazardous chemical pesticides.

Neelima Gupta and Varsha Gupta (eds.), *Experimental Protocols in Biotechnology*, Springer Protocols Handbooks,
https://doi.org/10.1007/978-1-0716-0607-0_5,

The animal-parasitic nematodes are of immense pathogenic significance in as much as they could also be of zoonotic significance, and occupy practically every organ of the body of a vertebrate (alimentary canal, coelom, liver, pancreas, lungs, brain, kidney, muscles, etc.), and have been reported most commonly from the coelom of a variety of arthropods, like cockroaches and *Tribolium*.

2 Technique

2.1 Plant- and Soil-Parasitic Nematodes

These are the microscopic organisms that are separated in water, from particles that settle faster and can subsequently be poured off (decanted). The separation of these minute organisms could be effectuated depending on their size, shape, and rate of migration through different sizes of sieve.

3 Materials

3.1 Plant-Parasitic Worms

It is a universal method involving mesh of different sizes (60, 100, 250, 350, and 400) and specific gravity of solution between nematodes and the soil components.

3.1.1 Extraction of Free-Living and Ectoparasitic Worms

Decanting and Sieving Technique

1. Mix the soil sample (500 g) into the bucket, half filled with water.
2. Allow heavy soil particles to settle down for a moment, decant through a 60 mesh sieve into another bucket.
3. The heavy silt particles would settle in about 10–20 sec, and decant the suspension onto a series of fine sieves (100, 250, 350, 400) where different sized nematodes (as per their body sizes) would be retained at every sieve.
4. Wash the suspension containing nematodes of each sieve and pour it through a fine wire gauze sieve containing double layered tissue paper already placed onto a beaker to retain the fine silt and leave the beaker overnight. The nematodes would pass through the minute gaps in the tissue paper and collect at the bottom of the beaker.
5. Finally reduce the contents of the beaker by passing through 400 mesh size sieve and pour the contents containing nematodes into cavity block and observe under binocular microscope.

Centrifugal Flotation Technique

In this technique sugar flotation solution of greater specific gravity (1.18) than that of the nematode (1.05) is generally used for the collection of nematodes at large numbers by adopting the following steps:

1. Obtain nematode suspension in the beaker after sieving as described in sieving and decanting technique.
2. Distribute the contents of beaker between 2 and 4 centrifuge tubes and centrifuged at 2250 × g for 4 min.
3. Nematodes and debris got compacted at the bottom of the centrifuge tube.
4. Pour off the supernatant with most of the debris with a rotating motion. Remove the leftover debris around lip of the tube with a clean fingertip.
5. Fill each tube half full with the sugar solution (dissolve 500 g of sugar in 1 L of water, add 10 mL of 10% lactic acid to the sugar solution to inhibit bacterial and mold growth).
6. Close the tube with a rubber stopper and shake until the compact mass at the bottom goes into suspension.

3.1.2 Extraction of Endoparasites

Dissecting Plant Materials in Water

1. Shred, with a dissecting needle, the infected plant material (e.g., angular spots manifesting leaves harboring foliar nematode, wheat seed galls infected with *Anguina tritici*, or root knots infected with *Meloidogyne* spp.) in water in a dish under dissecting microscope.
2. Collect the nematodes by a handling needle.

Modified Baermann Funnel Technique

Baermann funnel consists of the following materials:

1. A conical glass funnel (10 cm in diameter) (Fig. 1)
2. A short piece of rubber tubing attached to its stem and closed by a clamp.
3. A wire gauge sieve placed inside the top of the funnel, resting about 2 cm below the top rim.
4. A wet strength facial tissue, to retain roots or soil, is placed on the wire gauge sieve. Fill the funnel with tap water to a level above the wire gauge sieve.
   (a) Cut 10 g of fine rootlets/plant material into small pieces and wash.
   (b) Place the plant material onto a tissue paper and funnel is left as such for 2–5 days.
   (c) Collect the nematodes at the bottom of the funnel stem.
   (d) A small quantity of water containing the nematodes is run-off.

Mechanical Maceration Technique

1. Homogenize 5–10 g of clean roots or shoots in 50–100 mL water for 20 s.

Fig. 1 Extraction of nematode using Baermann funnel

2. Pour the suspension onto a tissue paper supported by a wire gauze sieve kept in a funnel holding sufficient water to remain in contact with the bottom of the wire gauze.
3. Leave funnel for 1–2 days.
4. Collect nematodes at the bottom of the funnel stem.

Root Incubation Technique

1. Thoroughly wash soil off roots with the water spray.
2. Place washed roots in polyethylene bags or glass jars where the atmosphere remains humid.
3. After 72 h flush the roots in the container thrice, onto a 325-mesh sieve held at an angle to extract nematodes, and pour into a 250 mL beaker.

3.1.3 Extraction of Cyst Nematodes

1. Drying of soil sample is essential.
2. Dry cysts float in water, whereas wet ones sink to the bottom and cannot be retrieved.
3. *Globodera* cysts—rounded to globulous, cyst wall variously patterned, often being zigzag or wavy lined or network; *Heterodera* cysts—pyriform or lemon shaped.

Handling of Nematodes

Picking and Transferring of Plant-Parasitic Nematodes

A peacock feather, sharpened to a very fine point fixed in a needle is used to pick up slender nematodes under the dissecting microscope. The nematode will be released from the pick in a small volume of water in a vial.

Simple Extraction Method

1. Place well mixed dried soil (50 g) in a 60-mesh sieve. Wash it.
2. The catch of the sieve is poured into a white bowl with a sloping edge.
3. The cysts float along the edge of the bowl and can be separated with a camel hair brush no. 1.

Acetone or Acetone Tetrachloride Method

1. Dry well mixed soil and place in 60mesh sieve.
2. Wash soil carefully and transfer the content into 250 mL volumetric flask.
3. Fill the flask (half) with acetone or acetone: carbon tetrachloride (3:1).
4. Shake the flask vigorously.
5. Fill the flask just below its tip and allow it to stand for 1 min.
6. Decant the floated cysts and debris through filter paper.
7. Place filter paper in a petri dish and view under microscope.

3.2 Animal-Parasitic Nematodes

The animal-parasitic nematodes are generally visible through naked eye and thus are removed from the bodies of their hosts manually, unless special techniques are to be employed.

3.2.1 Isolation Procedure

The alimentary canal of anesthetized vertebrate hosts is stretched after washing, kept in normal saline and dissected. The associated glands as well as other organs are also washed, kept in normal saline and dissected to free the entangled live roundworms within the body tissues. The worms are cleaned thoroughly with fresh normal saline to free these worms from debris or other digestive juices. Simultaneously, the live worms should be kept in German or other high-quality absolute alcohol for molecular work immediately after their extraction from body tissues.

3.2.2 Fixation of Worms

Kill the live worms in lukewarm water for morphological observations and taxometric analysis; and transfer to (hot) absolute alcohol–glycerin (95:5), keep overnight.

3.2.3 Preservation of Worms

Transfer the worms to lactophenol and keep these for current or future examination. Keep these worms on a slide under the cover glass for their morphometric examination under microscope.

Lactophenol

Lactic acid	100 mL	200,100
Phenol	100 mL	200,100
Distilled water	100 mL	200,100
Glycerol	100 mL	400,200

4 Methods

4.1 Killing and Fixation

The isolated nematodes are to be acclimatized for 18 h for permanent mounting. The following steps have to be carried out for permanent mounting of nematodes:

1. Manually pick nematode of similar appearance and body size, in the cavity blocks of 20 mm diameter, with the help of a fine needle mounted on a plastic holder.
2. Pour hot (at 70 °C) fixative in the cavity blocks containing nematodes. Killing by hot fixative has no distortion or twisting, facilitating better microscopic examination of minute structures like lateral pores, phasmids, and alae when mounted in glycerin.
3. Take care to avoid excess heating.
4. Cool the fixative at room temperature and remove the excess water with the help of a fine-tipped dropper.
5. Finally keep the nematodes in fixative for 48 h.

4.1.1 Fixative

Apply methods of Seinhorst [1] and De Grisse [2] for fixing the nematodes. De Grisse's [2] method is found to be more satisfactory, as the nematodes need to be transferred from one fixative to another. De Grisse's fixative also provides excellent view of the organs of the nematodes viz. oesophagus, intestine, reproductive organs etc. which have great taxonomic value [2]. The composition of the fixative is as follows:

Seinhorst's Fixative: F.A (4:1) [1]

Formalin 40%	4 ml
Glacial Acetic acid	1 ml
Distilled water	95 ml

De Grisse's Fixative: F.A.G (10:1:1) [2].

Formalin (40%)	10 ml
Glacial acetic acid	1 ml
Glycerin	1 ml
Distilled water	88 ml

4.1.2 Seinhorst's Method for mounting Nematodes

The following steps have to be carried out for nematode mounting [1]:

1. Transfer the previously killed nematodes with Seinhorst's fixative to 4% formalin in a cavity block and fix for 8–12 h, preferably overnight.
2. After overnight fixation transfer the nematodes to a mixture of ethanol (96%), glycerin, and distilled water (20:1:79). Cover the cavity block with a lid and place in a desiccator (saturated with ethanol vapor) in an incubator at 40 °C for about 12 h.
3. Take the cavity block out of the desiccator after 12 h. Draw the liquid out from the cavity to the maximum and replace with a mixture of ethanol (96%) and glycerin with ratio of 93:7 and cover with a cover glass lid.
4. Transfer the cavity block again into the incubator at 40 °C for 2–3 h. Thereafter add twice a drop of pure glycerin to the cavity block after every 1 hour. Cover the cavity block with glass lid and place in the incubator at 40 °C for another 2–3 h.
5. Lastly transfer the cavity block into a desiccator containing $CaCl_2$ and leave for overnight.
6. Now the nematodes are ready for mounting in pure glycerin.

4.1.3 De Grisse's Method for Mounting of Nematode [2]

Dehydrate the fixed nematodes for semipermanent mounting. The steps for making glycerin mount are as follows:

1. Fix the collected nematodes in De Grisse's fixative for about 48 h in a cavity block.
2. Reduce the fixative to 1 ml with the help of finely drawn pipette and keep at 40 °C in ethanol atmosphere for 6–8 h.
3. Remove the cavity block from ethanol atmosphere and add the fixative 2 (see composition), till the cavity block is completely filled-

Ethanol 96%	95 ml
Glycerin	5 ml

4. Cover the cavity block with glass lid and place it at 40 °C for 2–3 h.
5. Remove the lid of the cavity block after 2–3 h and allow the solution containing nematodes to evaporate.
6. Add fixative 3 (Seinhorst, [1]) with the following composition:

Ethanol 96%	50 ml
Glycerin	50 ml

7. Cover the cavity block and place it in an incubator at 40 °C.

8. After 1–2 h remove the lid of the cavity block and allow the solution to evaporate at 40 °C for nearly 2 h.
9. Finally place the cavity block in a desiccator containing $CaCl_2$ and leave overnight.
10. Now the nematodes are ready to mount in pure glycerin.

4.2 Preparation of Nematode Mounts

The following steps are to be carried out for mounting the nematodes for further study:

1. Put 4–5 similar nematodes of more or less of equal diameter to a small drop of glycerin on the slide and let the nematodes settle down at the bottom of the glycerin drop.
2. Put 3–4 glass wool rods of 0.25–0.5 cm long and almost of the same diameter in the glycerin drop in order to avoid the pressure of the mounting coverslip.
3. Place gently a coverslip over the drop of the glycerin.
4. Excess of glycerin flowing out of the coverslip be taken away by small pieces of blotting paper placing around the coverslip.
5. Seal the coverslip by applying glyceel.
6. Label the slide and nematodes are ready for microscopical examination.

4.2.1 Temporary Mounts

Place a drop of 3% formalin on a clean glass slide. Pick up 8 nematodes from the suspension and arrange them in 2 lines in formalin drop. Arrange 3 glass wool pieces radially around the nematodes. Apply a coverslip over the drop carefully. Drain off the excess fluid. Seal the edge of the coverslip with a ringing medium like "Zut" or "Glyceel." Label the slide.

4.2.2 Semipermanent Mounts

Collect the nematodes, kill and fix them as described in the above method. Remove the nematode into a drop of warm, vaporizing lactophenol and allow them to cool. Fix the nematodes in a small drop of lactophenol on a final glass slide. Place a coverslip with a ringing medium like nail polish, "Zut," or "Glyceel." Label the slide.

4.2.3 Permanent Mounts

Lactophenol Method

1. Take lactophenol in a cavity slide containing 0.01% cotton blue stain and heat on a hot plate to about 65 °C.
2. Transfer already fixed nematodes in hot lactophenol onto a slide and examine under a stereoscopic microscope.

Glycerol Method

1. Transfer fixed nematodes to about 2 ml glycerol (1.5%; with trace of $CuSO_4$ or thymol to prevent the growth of molds) in a cavity block.

2. Keep the cavity block air-tight together with a small tube of desiccant such as ($CaCl_2$ or silica gel) at 25–30 °C for 4 weeks.
3. Place a drop of dehydrated glycerol on a clean slide with 8 nematodes from the dish.
4. Arrange 3 glass wool pieces radially and apply a coverslip over the drop carefully. Seal the edge of the coverslip with "Glyceel."

4.3 Molecular Investigations

4.3.1 Genomic DNA Extraction

First, take the tissue and crush it. The DNA isolation shall be performed by using the DNeasy Blood and Tissue Kit (Qiagen) by performing the following steps.

1. Transfer surface sterilized Tissue/IJs into a 1.5 ml centrifuge tube.
2. Add 200 μl lysis buffer (ATL) and crush tissue/IJs with mortar and pestle or tissue homogenizer.
3. Add 20 μl proteinase K in the sample and incubate in water bath at 56 °C for 3 h or till it is completely lysed. Vortex occasionally during incubation.
4. Add 200 μl Buffer AL and mix thoroughly by vortexing. Again, incubate it at 56 °C for 10–15 min.
5. Add 200 μl of ethanol and mix properly to make homogeneous solution.
6. Transfer the mixture into mini spin column placed in a 2 ml collection tube and centrifuge at 6000 × g for 1 min. Discard the supernatant and collection tube.
7. Attach a new collection tube to mini spin column and add 500 μl of prewash buffer AW1 to the mini spin column. Centrifuge at 6000 × g for 1 min and discard the collection tube and supernatant.
8. Place the spin column in a new 2 ml collection tube. Add 500 μl wash buffer 2 (AW2) in mini column and centrifuge at 9000 × g for 3 min. Discard the supernatant along with collection tube.
9. Transfer the spin column to a new 1.5 ml microcentrifuge tube and add 200 μl of elution buffer AE at the center of spin column membrane and avoid splitting on the sides. Incubate it for 5 min at room temperature and centrifuge at 6000 × g for 2 min. Store the eluted genomic DNA at −20 °C for the further use.

4.3.2 Agarose Gel Electrophoresis

1. Weigh 0.375 g of agarose and dissolve it in 50 ml of 1× TAE (Tris acetate EDTA) buffer in microwave or by heat in a conical flask.

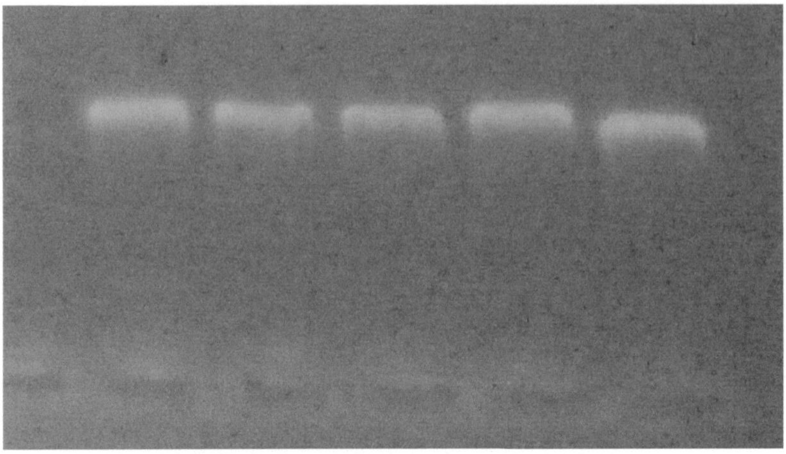

Fig. 2 Visualization of DNA under UV transilluminator

2. Cool till it becomes bearable to touch and add 1 μl of ethidium bromide (EtBr). Mix and pour it into gel casting tray with comb attached, leave till it solidifies.
3. Remove the comb and place the casting tray with the gel in electrophoretic tank.
4. Mix 3–5 μl DNA template with 1 μl of DNA loading dye (containing bromophenol blue, glycerol, and sucrose) and load it in a well of the gel. Repeat the same with other samples and load them individually in the corresponding wells.
5. Run the electrophoretic unit filled with 1× TAE at 100 V, 400 A for 30 min and observe the gel under UV transilluminator for the presence of DNA (Fig. 2).

4.3.3 PCR Amplification

The genomic DNA will be amplified using the 96-well fast thermal cycler. Different regions of taxonomic importance like ITS rRNA, 18S rRNA, 28S rRNA, and cytochrome oxidase subunit I will be amplified using primers given below:

ITS1	SS1-GTTTCCGTAGGTGAACCTGCG-
	NC13R-GCTGCGTTCTTCATCGAT-
NEM 18S	18S F-CGCGAATRGCTCATTACAACAGC-
	18S R (5′-GGGCGGTATCTGATCGCC-3-
18S	SSU18A (AAAGATTAAGCCATGCATG-
	SSU26R (CATTCTTGGCAAATGCTTTCG-
COI	SyphaCOIF 5′-GGTCTGGTTTTGTTGGTAGTT-3′
	SyphaCOIR 5′-AACCACCCAACGTAAAC ATAAA-3′

The PCR reaction mixture (25 μl) consists of the following composition:

Forward Primer-	1 μl
Reverse Primer-	1 μl
DNA Sample-	5 μl
Nuclease Free DW-	8 μl
PCR Master Mix-	10 μl

The cycling profile will include initial temperature 94 °C for 5 min, 35 cycles of denaturation at 94 °C for 45 s, annealing temperature (as mentioned in the primer list, different for different primers), primer extension 72 °C for 1 min, final extension 72 °C for 8 min and storage temperature 4 °C for infinity (Fig. 3).

4.3.4 DNA Barcoding

Analyze and sequence the amplified PCR Product of DNA. Edit the DNA sequences and generate the contig sequences through DNA Baser version 4.20.0 and Bio-Edit [3] (Fig. 4). These shall be submitted to National Centre for Biotechnology Information (NCBI) through BankIt submission tool. The sequences are to be compared by using computational technique, BLAST (Basic Local Alignment Search Tools) with the other related species sequences available at the GenBank of NCBI (National Centre for Biotechnology Information) (Fig. 5). The phylogenetic analysis of all the sequences will be performed using software MEGA 7.0 [4]. Available data are retrieved from the NCBI database and utilized for the construction of distance matrix and phylogenetic analysis through maximum parsimony method (Fig. 6). The sequence of *Caenorhabditis elegans* is taken and used as an outgroup for generating the trees (Fig. 6).

4.4 Some Important Nematodes Parasitizing Animals

4.4.1 Scanning Electron Microscopic Analysis

Scanning Electron microscopy is the specialized technique that has helped specifically to bring prominent morphological features of helminthes. This is of significant taxonomic significance, to the knowledge of taxonomists and has opened up horizons to resolve taxonomic riddles that created confusions due to overlapping features of nematodes amongst different families.

In Invertebrate Hosts

The worms of *Leidynema capoori* isolated from cockroach *Periplaneta americana* were subjected to morphometric analysis as well as SEM analysis to fetch out detailed taxonomic characteristics. The vestibule of buccal cavity is surrounded by a thickened rim comprised of 12 small laminated flaps, that overlapped one over the other and locked together in a sequence with margins visible distinctly. These laminated structures apparently contributed to the formation of a thinner circular lid over the vestibule of buccal cavity. Body robust, spindle shaped, cylindrical, elongate, wide, maximum diameter at the level of vulva, with a terminal spine and a relatively

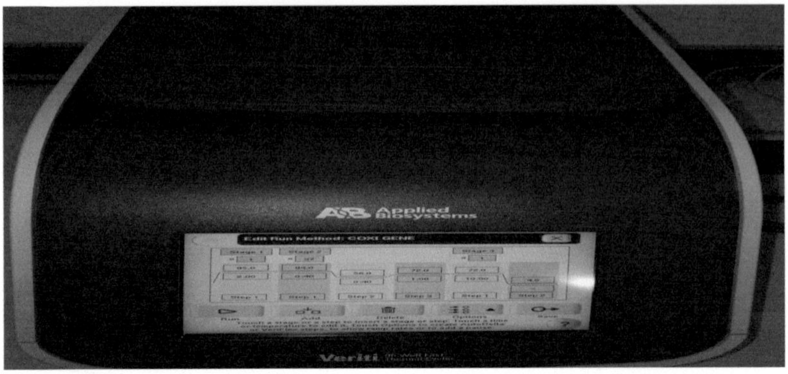

Fig. 3 PCR cycling profile

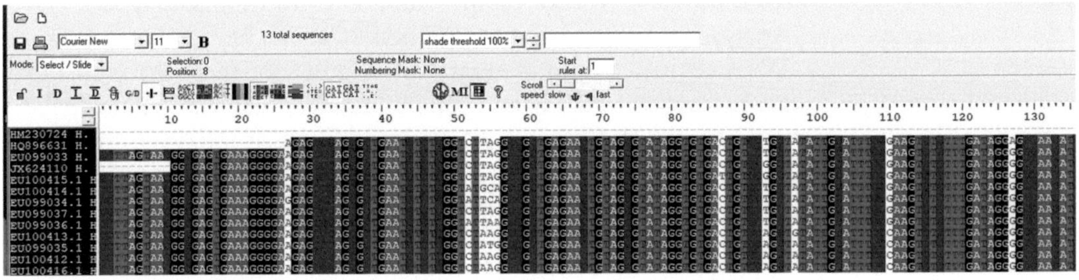

Fig. 4 Edited sequences after multiple sequence alignment (MSA) in BioEdit

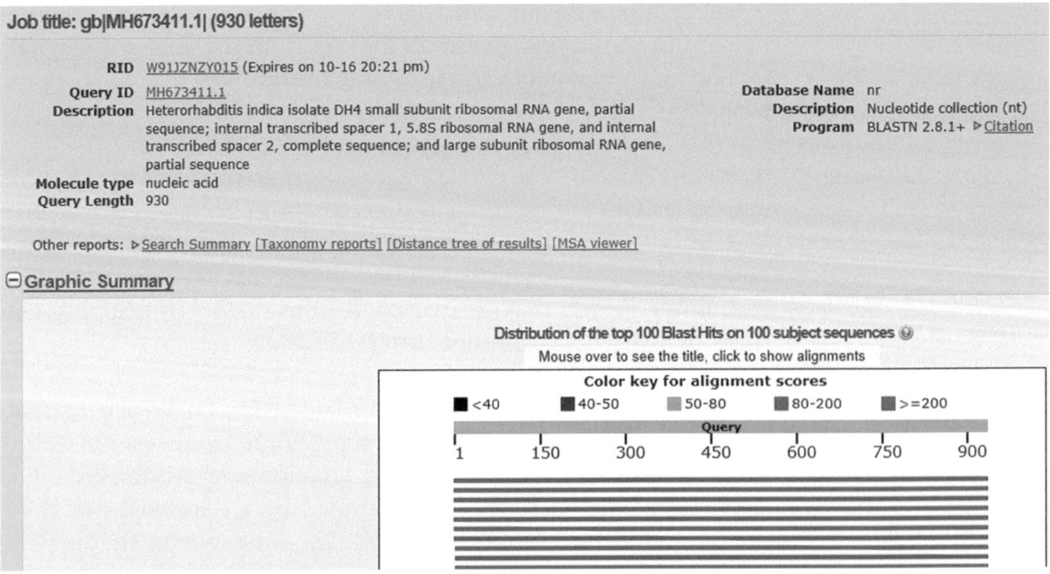

Fig. 5 Summary of Basic Local Alignment Tool (BLAST)

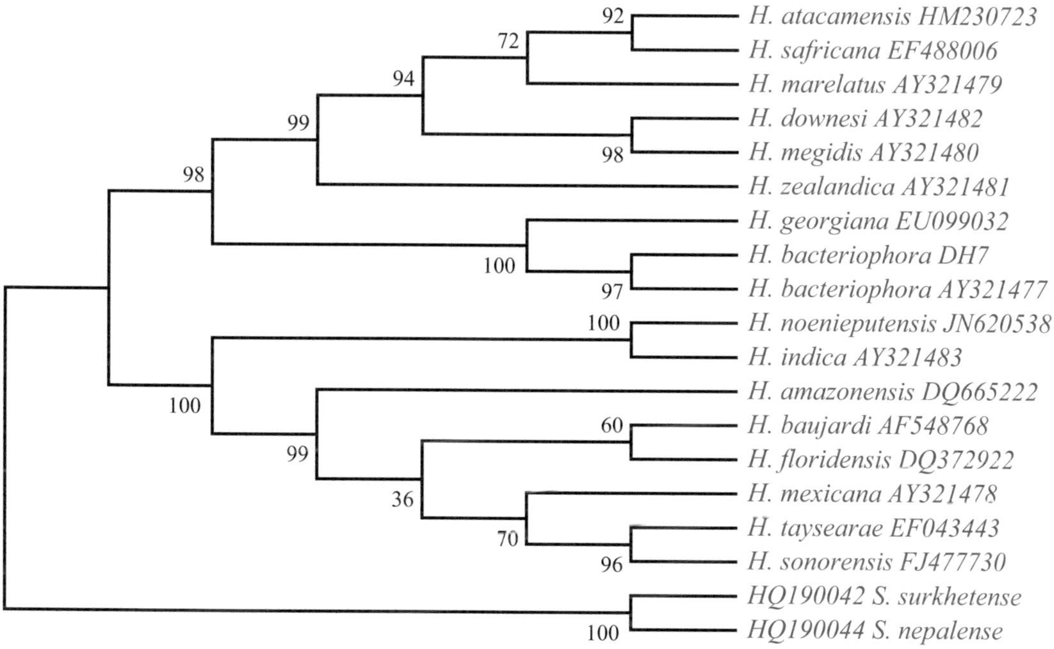

Fig. 6 Evolutionary relationships of taxa by maximum parsimony method

longer tail. The base of circular rim comprising lips is supported by 15–16 girdle-like supporting rods. The spaces between annulations were traversed by thickened extensions at the lateral margins, whereas the space between front oral end and the first annulations comprised. Cuticular annulations prominent, with marked annulations between cephalic end and the base of procorpus.

In Vertebrate Aquatic Hosts Occurring in Freshwater Habitats

As extracted during dissection—The nematodes isolated after Habitats dissection of intestine and other internal organs of Bagarius bagarius and Duttaphrynus melanostictus were washed in warm water and fixed in hot 96% alcohol + Glycerine (95:5). After 24 hrs. worms were kept in Lactophenol for having these photographed and presented in Figs. 7 and 8. *Indospinezia spinicaudatum* were from *Xenentodon cancila* inhabiting River Yamuna at Allahabad (Figs. 9 and 10a)

In Vertebrate Aquatic Hosts Occurring in Marine and Freshwater Habitats

SEM studies illustratively highlighted comparable morphological features between the roundworms of mysterious invasive potential, *Rostellascaris spinicaudatum* were recovered from *Mystus seenghala* and *Bagarius bagarius* of River Ganges at Allahabad and marine fishes *Rhynchodon typus* and *Arius falcarius* of central west coast of India at Goa.

The occurrence of unique sunflower preanal papillae (Fig. 10b) in the anisakid roundworms encountered in fishes from river Ganges, including the fishes inhabiting the tributaries of River Ganges that enter into Bangladesh (*Goezia bangladeshi*), in a

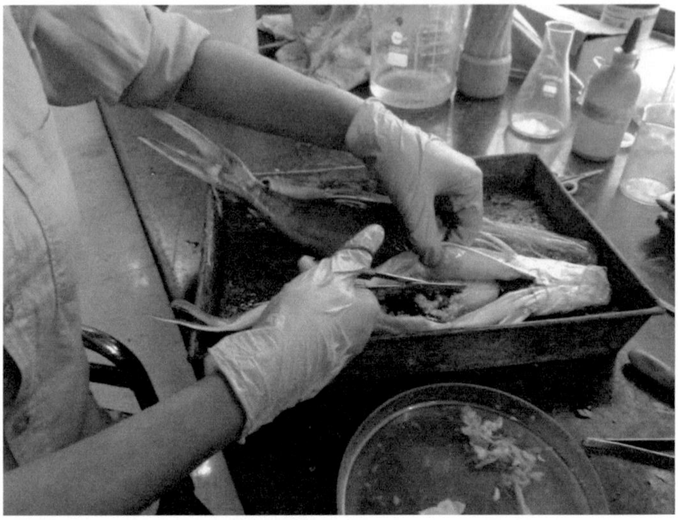

Fig 7. Bagarius bagarius being dissected

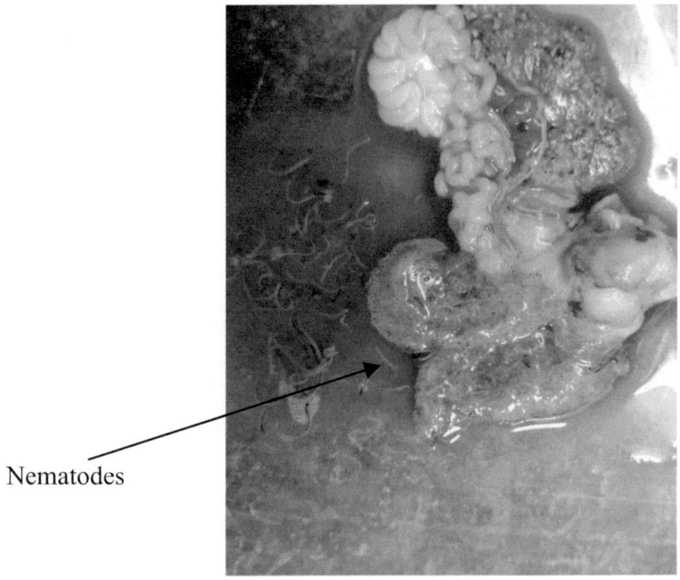

Fig 8. Neyraplectana in Duttaphrynus melanostictus

frequent manner could be successfully established with the aid of SEM analysis. The freshly collected live worms from fishes were fixed in 4% glutaraldehyde for further processing to obtain images of SEM.

The morphology of head (Fig. 10c) of anisakids, raphidascaridoids (Fig. 10a) and the lastomatoid worms inhabiting cockroaches (Fig. 11a, b) distinctly elaborating structures of vestibule

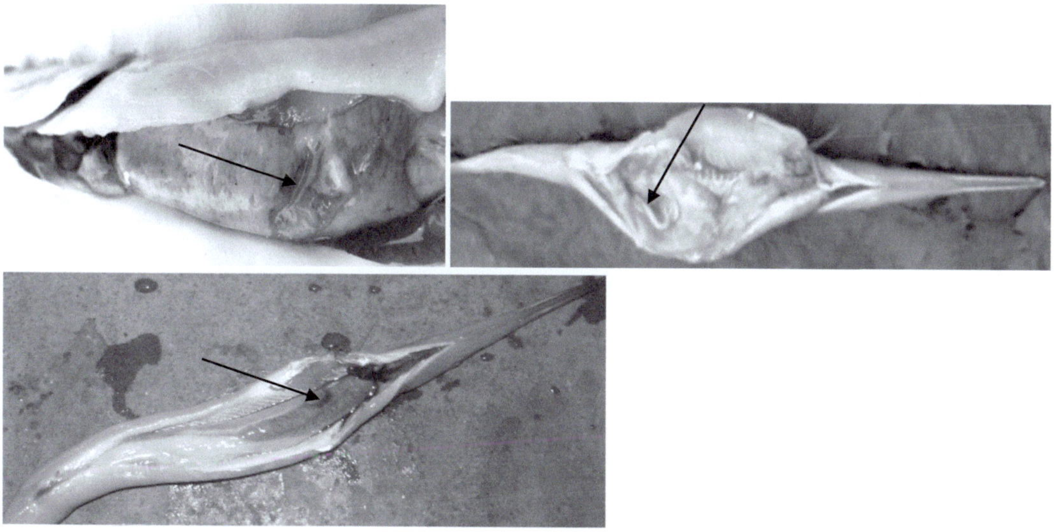

Fig. 9 *Xcnentodon cancila—Indospinezia multispinatum* in liver

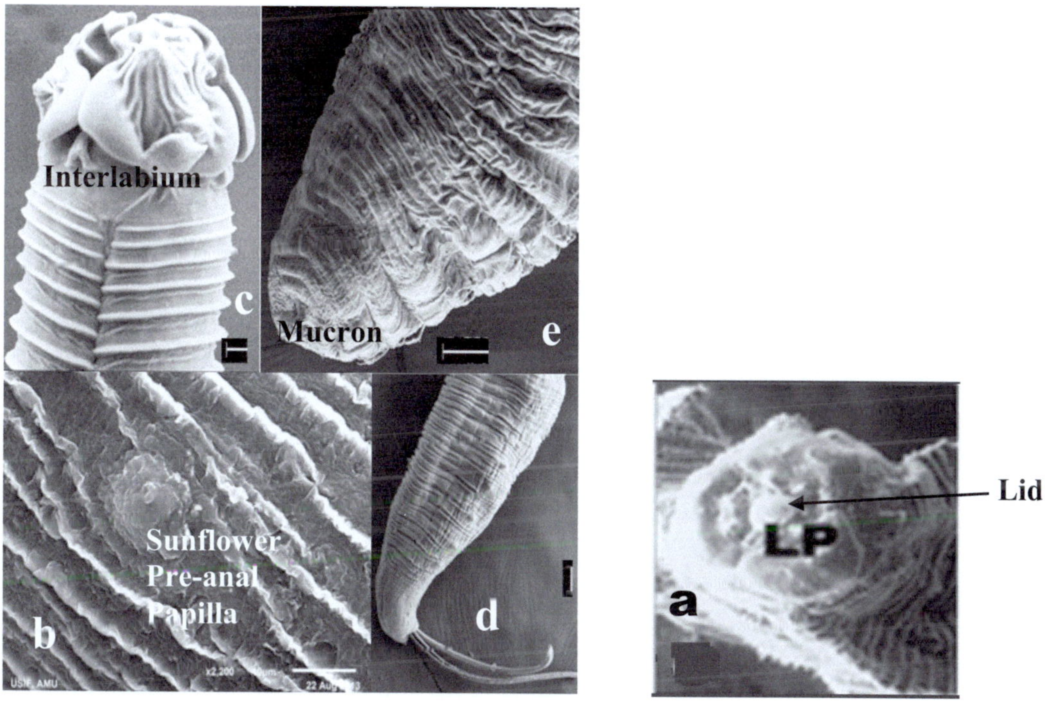

Fig. 10 (a) *Indospinezia multispinatum*—en face *Bagarius bagarius* of River Ganges at Allahabad.view: SEM to show lid-plate atop oral end of male worm. (*LP* Lid-plate) (b–e) from *Rostellascaris spinicaudatum*

of buccal cavity and lips around it could specifically be illustrated using SEM technique. The similarity in structure of the lid covering buccal cavity on oral surface of a raphidascaridoid (Fig. 10a) and the lastomatoid worm (Fig. 11b) is a unique structural peculiarity,

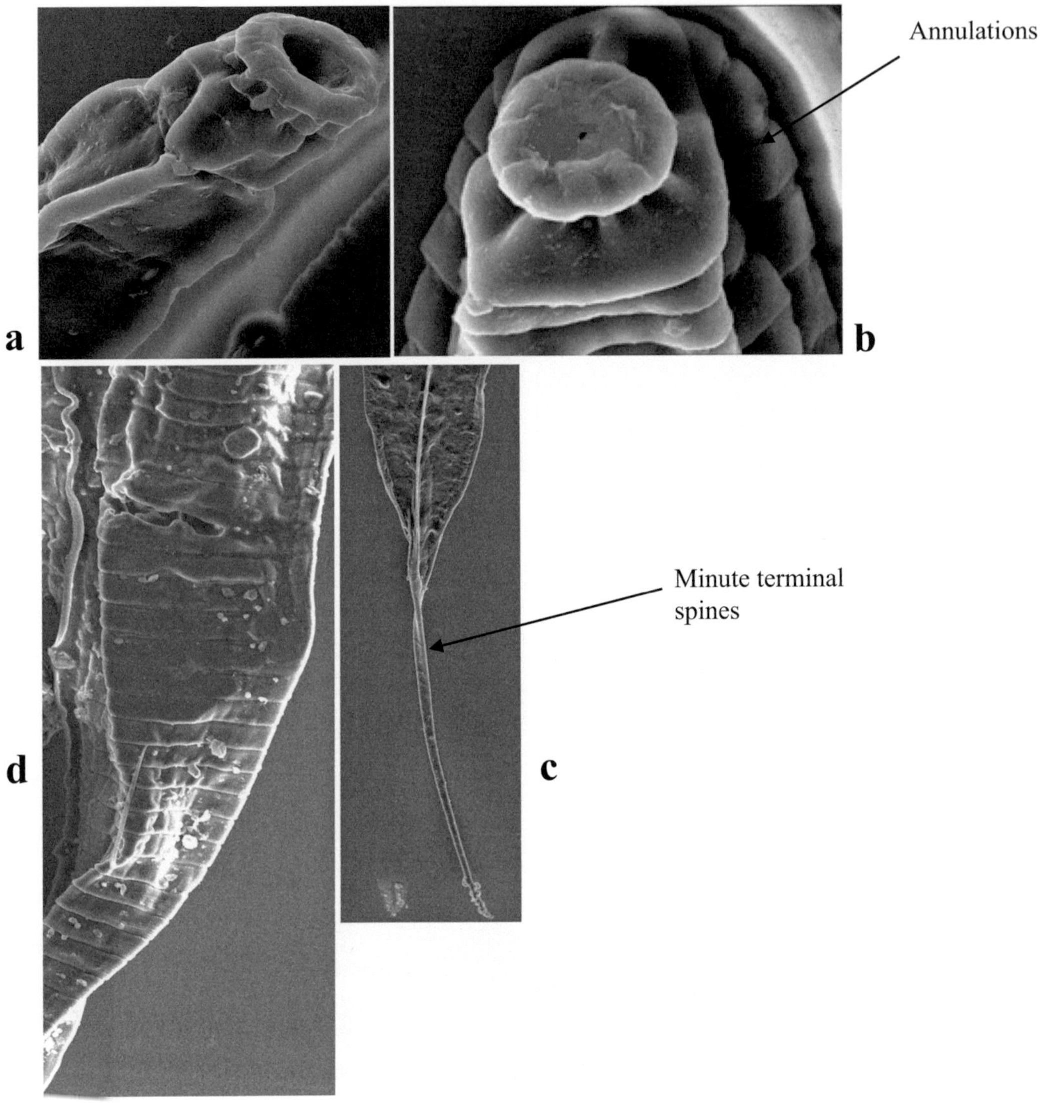

Fig. 11 (**a**–**d**) *Leidynema capoori*—anterior end of female to show buccal cavity (**a**); oral end of female to show lid-plate over buccal cavity (**b**); posterior end of female to show minute spine on tail tip (**c**); posterior end of male to show caudal alae and post-anal papillae (**d**)

hitherto unknown, but revealed by SEM. The significant variations in the difference in morphology of tail of nematodes could be highlighted by the use of SEM studies. The blunt mucron at the tail tip of *R. spinicaudatum* (Fig. 10d, e) is distinguishable from the spiny tail tip of *Leidynema capoori* (Fig. 11c) and the differences in tails of female and male *Leidynema capoori* (Fig. 11c, d) as revealed by SEM.

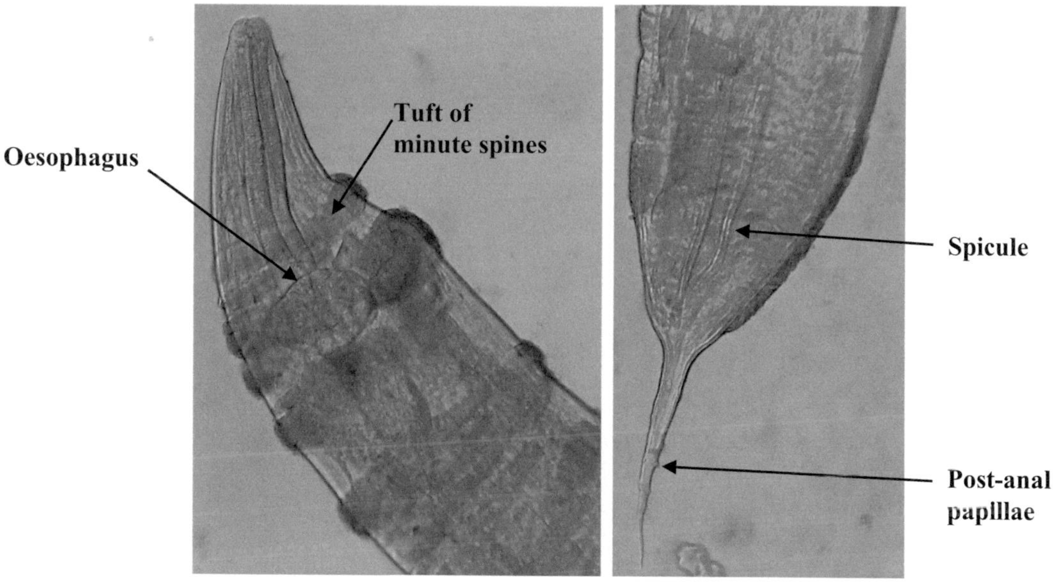

Fig. 12 *Aplectana* sp. Male from *Duttaphrynus melanostictus* at Allahabad University campus

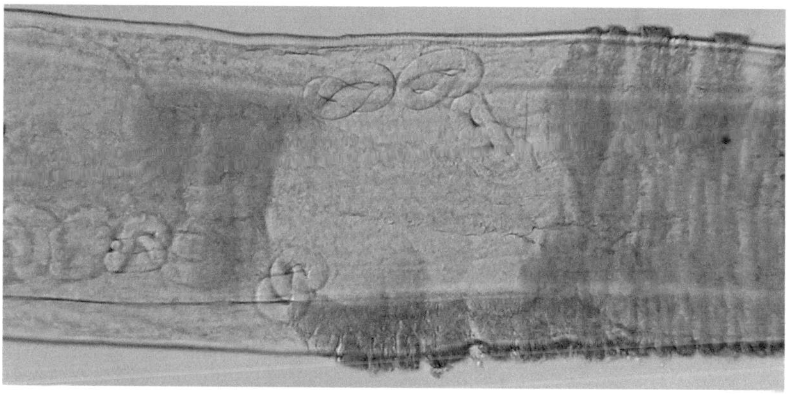

Fig. 13 *Aplectana* sp. Female viviparity

In Amphibian Hosts

The various morphological features of the worms of both sexes of *Aplectana* sp. from *Duttaphrynus melanostictus* at Allahabad University campus have been reported (Figs. 12, 13 and 14).

In Avian Hosts

A variety of nematodes, *Heterakis* sp., and cestodes, *Raillietina* (*Raillietina*) sp. parasitizing avian hosts (*Columba livia domestica*), have been documented (Fig. 15).

4.5 Molecular Taxonomy Procedures

Worms are used for gene sequencing of ITS gene, ITS1, ITS2, 28S rRNA, 18S rRNA [5] and mit coi genes to provide evidence of taxa segregation from closely related species, genera or other higher taxa on the basis of phylogenetic analysis through construction of

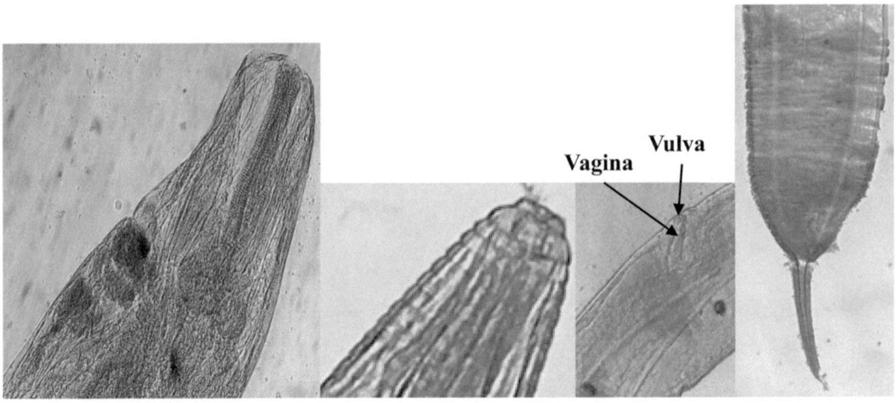

Fig. 14 *Aplectana sp.* Female (anterior end, vulva & posterior end) from *Duttaphrynus melanostictus* at Allahabad University campus

Fig. 15 Cestodes, *Raillietina* (*Raillietina*) sp., and Nematodes, *Heterakis* sp., as dissected and extracted out of small intestine of pigeon (*Columba livia domestica*)

neighbor-joining tree or maximum parsimony tree. The nematodes are freshly washed with warm water; kept in normal saline during the period of their extraction from the body of their hosts, and processed immediately after transferring these to absolute alcohol [6].

The total genomic DNA is isolated from the nematode samples collected from *D. melanostictus* and *H. tigerinus* by using Qiagen DNeasy Blood and Tissue Kit (Qiagen, USA), following the manufacturer's instructions. The D2D3 LSU expansion segment of 28 S ribosomal RNA gene is amplified using the universal primers D2A (5′AGCGGAGGAAAAGAAACTAA3′) and D2R (5′TCGGAAGGAACCAGCTACTA3′). The thermal regime consists of an initial denaturation at 94 °C for 3 min followed by 30 cycles at 94 °C for 30 s, 50 °C for 30 s, and 72 °C for 1 min, and a final extension for 10 min at 72 °C. The rDNA regions comprising ITS and ITS2 sequences are amplified with primers, 18SF (5′TTGATTAGGTCCCTGCCCTTT3′) and 26SR (5′TTTCACT CGCCGTTACTAAGG3′) for ITS gene, and SS2 (5′TTGCAGACACATTGAGCACT3′) and NC2 (5′TTAGTTT CTTTTCCGCT3′) for ITS2 gene. Each PCR reaction is performed under the following conditions: after initial denaturation at 94 °C for 5 min, 35 cycles of 94 °C for 30 s (denaturation), 50 °C for 30 s (annealing), and 72 °C for 30 s (extension), followed by a final extension at 72 °C for 8 min. The PCR products are run on a 1% agarose gel and visualized by ethidium bromide staining. The primers used for the amplification of the coi gene were forward primer LCO1490, 50-GGTCAACAAATCATAAAGATATTGG-30; reverse primer HCO2198, 50-TAAACTTCAGGGTGACCAAAAAATCA-30 [7]. The thermal regime consisted of an initial denaturation at 94 °C for 1 min, followed by 30 cycles at 94 °C for 1 min, 48 °C for 1 min, 72 °C for 2 min, and a postamplification extension for 7 min at 72 °C. The small subunit ribosomal RNA (18S rRNA) gene is also amplified in 50 μl volume with 5 μl of 10X Taq polymerase buffer, 2 ml of $MgCl_2$ (25 mM), 0.25 μl of each dNTP (0.05 mM), 0.5 μl of each primer (0.01 mM), 0.6 U of Taq polymerase, and 2 μl of genomic DNA. The primers used for the amplification of the 18S rRNA gene are Nem18SF, 50-CGCGAATRGCTCATTACAACAGC-30; and Nem18SR, 50-GGGCGGTATCTGATCGCC-30 [5]. The thermal regime consisted of an initial denaturation step of 5 min at 94 °C followed by 35 cycles of 30 s at 94 °C, 30 s at 54 °C, and 1 min at 72 °C, followed, in turn, by a final extension of 10 min at 72 °C.

After sequencing, DNA sequences are aligned for the phylogenetic analysis using the CLUSTAL W computer program [8], and DNA sequences are edited in BIOEDIT [3]. The phylogenetic trees are constructed by the neighbor-joining method [9] using MEGA version 6.0 [10]. The trees are evaluated using the bootstrap test [11] based on 1000 replications.

5 Notes

The researcher shall understand the implications of life cycle and habitats of plant-parasitic and animal-parasitic roundworms. The researcher should understand the significance of precise methods of nematode isolation for their ultimate use in molecular studies.

For identification minute details on the bodily organization of nematodes by the application of SEM will help to elaborate differentiating characters on the basis of attributes of taxonomic significance.

The elements of molecular taxonomy would pave the way for segregation of invasive species from those existing in their native habitats, besides providing reliable taxonomic criteria to differentiate closer species and other taxa.

References

1. Seinhorst JW (1962) Extraction methods for nematodes inhabiting soil. In: Murphy PW (Ed.), Progress in Soil Zoology, Butterworths, London 243–256
2. De Grisse AT (1969) Redescription ou modifications de quelques technique utilisées dans l'étude des nématodes phytoparasitaires. Meded. Fac. Landb Wettens 34: 351–369
3. Hall TA (1999) BioEdit: a user-friendly biological sequence alignment editor and analysis program for windows 95/98/NT. Nucleic Acids Symp Ser 41:95–98
4. Kumar S, Stecher G, Tamura K (2016) MEGA7: molecular evolutionary genetics analysis version 7.0 for bigger datasets. Mol Biol Evol 33:1870–1874. https://doi.org/10.1093/molbev/msw054
5. Floyd RM, Rogers AD, Lambshead PJD, Smith CR (2005) Nematode-specific PCR primers for the 18S small subunit rRNA gene. Mol Ecol Notes 5:611–612
6. Sandeep MK (1986) Bioecology of the parasites of high altitude homeothermic host–parasite systems. I. Influence of season and temperature on infection by strobilocerci of three species of Hydatigera in Indian rat host. J Helminthol 60:15–20
7. Folmer O, Black M, Hoeh W, Lutz R, Vrijenhoek R (1994) DNA primers for amplification of mitochondrial cytochrome c oxidase subunit I from diverse metazoan invertebrates. Mol Mar Biol Biotechnol 3:294–299
8. Thompson JD, Gibson TJ, Plewniak F et al (1997) The CLUSTAL_X windows interface: flexible strategies for multiple sequence alignment aided by quality analysis tools. Nucleic Acids Res 25(24):4876–4882
9. Saitou N, Nei M (1987) The neighbor-joining method: a new method for reconstructing phylogenetic trees. Mol Biol Evol 4:406–425
10. Tamura K, Peterson D, Peterson N et al (2011) MEGA5: molecular evolutionary genetics analysis using maximum likelihood, evolutionary distance, and maximum parsimony methods. Mol Biol Evol 28(10):2731–2739
11. Felsenstein J (1985) Confidence limits on phylogenies: an approach using the bootstrap. Evolution 39:783–791

Chapter 6

Computational Approaches in Drug Designing and Their Applications

Dev Bukhsh Singh and Rajesh Kumar Pathak

Abstract

Computational approaches have tremendous potential to speed up the process of drug discovery. There are several tools based on the methods and algorithms of computer science which can be used for structure modeling, cavity/binding site prediction, molecular docking and virtual screening, visualization and interaction analysis of docked structure, pharmacophore mapping, de novo ligand designing, lead optimization, molecular dynamics simulation, predicting absorption, distribution, metabolism, excretion, and toxicity (ADMET), and building quantitative structure–activity (QSAR) model. The selection of a potential drug target, its structural features, and precise information of the binding site is a very important step to proceed for drug discovery and designing. Target–ligand complex analysis and pharmacophore mapping can guide the process of de novo drug designing. Molecular docking and simulation analysis can be used to evaluate the stability of a ligand in the binding site of the target and also provides information about the residues involved in the interaction. Pharmacokinetics and pharmacodynamics optimization of lead compounds are required to improve the specificity, affinity for binding to the target, and their absorption, distribution to a target site, metabolism of drug, excretion, and toxicity. Prior ADMET prediction using different models can reduce the risk of drug failure in a clinical trial. QSAR model can be developed for the biological activity of known drugs against a target, which can be used for predicting the biological activity of a new molecule. These computer-aided drug designing (CADD) techniques are more efficient, less costly, and less time-consuming as compared to the traditional methods of drug discovery. These CADD approaches/tools have been designed by incorporating the parameter of basics scientific principles related to the problem, and prediction models are derived and validated by vast dataset. Still, there are some limitations related to different CADD tools, which may be overcome by including some other relevant parameters and dataset in a program, and also by increasing the computational capability.

Keywords Computer-aided drug designing, Molecular modeling, Lead optimization, Pharmacophore mapping, ADMET, QSAR

1 Introduction

There are many drugs which are in use for the treatment of different diseases for a long time. However, due to development of resistance now these drugs are not very effective against disease. There are many bacterial, viral, and fungal infections against which existing drugs are not highly efficacious. Gene encoding for a drug target

Neelima Gupta and Varsha Gupta (eds.), *Experimental Protocols in Biotechnology*, Springer Protocols Handbooks,
https://doi.org/10.1007/978-1-0716-0607-0_6,

may undergo mutation, which results in structural changes in the 3D conformation of target. Because of this mutation, the drug cannot effectively bind with the target and is not able to generate the desired effect. The target protein can be isolated from the patient, and its 3D structure can be determined or modeled. The binding site and amino acids of drug targets that are in interaction with a drug can be determined from the complex structures obtained by X-ray/NMR study.

Drug designing tools can be used to develop a more potent inhibitor for a target. Many cases of drug resistance have been reported. The computer software can be used to analyze the interaction and binding of drugs with the target enzyme. There are many targets available for the treatment of the same disease. The selection of an appropriate target for drug development is the most important and critical point. The drug target should be unique, specific, and should essentially generate a therapeutic response. The interaction of the drug with a target can be compared using structure analysis tools. In the case of drug resistance, the structure of a drug molecule can be modified keeping in mind the structural changes that have occurred in the mutant. In this way, many potential drug candidates against a target can be designed using drug designing software's which can be further tested in the in vitro and in vivo systems. During drug designing, emphasis should also be given to increase interaction between the drug and target. Binding affinity or interactions can be improved by adding a ring or group, removing a ring or group, substituting a ring or group with another ring or group. The drug design software can screen the compounds having a favorable and high affinity for drug targets using theoretical models of energy calculations.

These new molecules can be better than earlier known drugs. But there is a need to evaluate the absorption, distribution, metabolism, excretion, and toxicity of these new drug molecules. Software models are also available for these predictions. Many researchers are working in the direction of drug resistance to find a more potent drug which can suit well with the mutant target. We hope that such types of computational approaches and tools can generate better results in the field of drug-resistance therapy.

2 Technique

2.1 Computer-Aided Drug Designing

There must be a better understanding of the disease and related mechanism to proceed for drug designing. For structure-based drug design, the molecular structure of the target is the first and most requisite condition, which can be either determined by the experimental techniques or can be modeled using the structure of similar proteins [1]. Steps involved in computer-aided drug discovery and designing are shown in Fig. 1. The detailed information

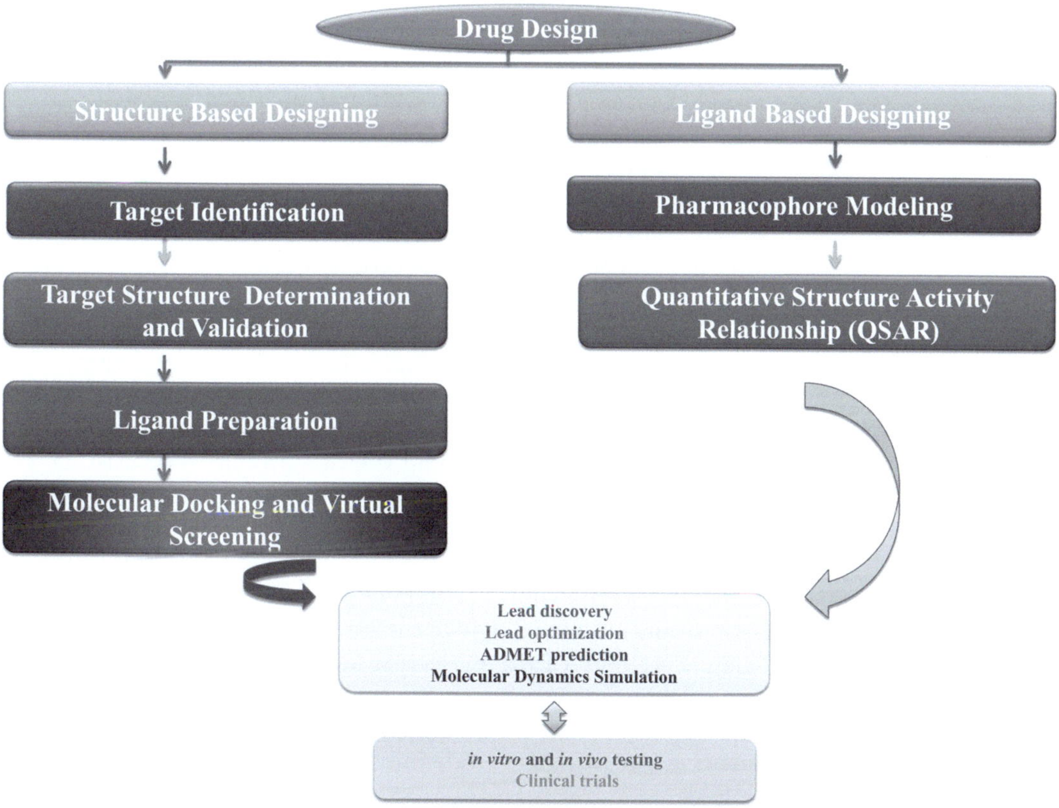

Fig. 1 Steps involved in drug discovery and designing through computational and experimental approaches

about the binding site of the target is required to understand the interaction. Molecular docking software provides the theoretical calculation of binding affinity and also the information related to the binding mode and interacting amino acids in a protein–ligand complex. Small molecule databases are very helpful in providing a vast amount of chemical structure diversity and scaffold. Lead compounds for drug discovery can be identified from the known natural bioactive molecules against disease, and virtual screening can also find some leads which can be optimized to have better pharmacokinetics and pharmacodynamics. A series of chemical modifications in a lead compound can be carried out to improve the binding affinity, specificity, absorption, distribution, metabolism, excretion, and toxicity. In addition to the above, quantitative structure–activity relationship (QSAR) methods can be applied to predict the biological activity of a compound from its physicochemical properties.

3 Computational Approaches

Several computational approaches and methods are available for drug designing. The development of database resources related to protein sequence and its 3D structure and drugs like chemical compounds and natural bioactive molecules have accelerated the process of drug designing [2]. A summary of different computational approaches and their application in drug designing have been discussed here.

3.1 Molecular Modeling

The 3D structure of a target is very important for understanding the role of the target in the progression of disease or therapy. The target structure can decipher the dynamics of the target protein and also the dynamic changes that take place during the binding of a ligand or drug with the binding site of a protein. The existing 3D structure of proteins and analysis tools provide a great extent of information related to the protein structure, which can be used for modeling the structure of new protein. Intramolecular or intermolecular forces stabilizing a protein or protein–ligand complex can be theoretically calculated, which can be used to interpret the thermodynamic stability of the molecular system. Several tools/software are available for protein modeling, which are based on the three basic approaches of modeling, that is, homology modeling also known as comparative modeling, fold recognition, and ab initio method.

3.1.1 Homology Modeling

This method models the 3D structure of a protein from its sequence information, that is, the amino acid sequence. This method assumes that proteins have an evolutionary connection, and the two or more proteins with a good extent of similarity can have a similar structure. In this way, the structure of a target protein can be modeled using the 3D structure of a known template. This method can be used only if the identity between target and template sequence is greater than 30%. Higher the identity between target and template more will be the accuracy of the modeled structure.

Homology modeling consists of the following steps:

1. Identify or retrieve the sequence of amino acids in a target protein.
2. Use the target sequence for searching the template from the database based on similarity/identity.
3. Select a template protein for modeling the target.
4. Align the target sequence with template structures to find the structurally conserved region (SCR), loop region, and amino acid residues varying in the side chain.

5. Assign the coordinate of all SCRs from the template to the target.
6. Search for the loop structure or generate the loop for the residues corresponding to the loop region.
7. Assign the coordinate of a loop at the right place in between two SCRs of the target.
8. Identify the residues/position whose side chains vary in the SCRs; here coordinates of backbone for such residues have already been assigned to the target.
9. From rotamer library, the bioactive conformation of the right-side chain is assigned/placed to the template.
10. Evaluate the accuracy of modeled structure using different available tools.
11. Refine the structure by reducing bumping/steric clashes between atoms or by minimizing the energy.

3.1.2 Fold Recognition

Homology modeling may not generate a reliable structure whenever the similarity between target and template is less than 30% [3]. Therefore, the threading or fold recognition method is employed for building a reliable model, and this approach includes more detailed structural information in predicting the best fold structure for a protein sequence. For example, the 3D profile method utilizes the knowledge of the environmental class of different residues in a protein. Structural analysis of proteins has shown that each amino acid has a different tendency to stay in a particular environmental class. There are six basic environmental classes, and a total of 18 environmental classes for each residue in three secondary structures, that is, helix, sheet, and coil, exist. The basic steps involved in fold recognition are given below.

1. Search for all possible folds of a target sequence from structure databases.
2. Compare the target sequence with all possible folds.
3. Evaluate the compatibility or fitness of folds with its target sequence.
4. In the 3D profile method, each 3D fold is converted into a 1D array of environmental class, and then the score of each fold is calculated by aligning/comparing the a 1D array of the target sequence with 1D array of an environmental class of fold. Here, we evaluate the compatibility of each residue in target with its environmental class in the given fold.

3.1.3 *Ab Initio* Approach

This approach is used when no structural counterparts or structural fold for a target protein is available. This method generates a protein structure by determining the configuration space of atoms in an amino acid using the knowledge of various principles of physics, chemistry, and mathematics. This approach requires a search method to explore the entire conformational space and an energy-based scoring function to find the accurate structure. Different conformational space can be generated by allowing the amino acids to assume the different discrete values of torsion angels. Energy for a conformation is calculated using the appropriate constraints such as a number of bonds, bond length, interatomic distances, disulfide linkage, and so on. A structure with global minimum energy conformation is considered to be a stable and native form.

3.1.4 Model Optimization and Validation

An accurate side chain conformation (rotamer) is required for better packing with the backbone. There might be steric clashes between charged atoms as well as bumping due to wrong placement of an atom position. All these factors cause disorder in the system due to the increase in energy of the structure. Model optimization requires a more accurate energy function for force field calculation. Accuracy of the modeled structure can be validated using different tools and servers such as Procheck, Whatcheck, Errat, and Verify3D. Normality indices can be used to compare the feature in a model that resembles with the structures determined by the X-ray crystallography and NMR spectroscopy. Many characteristics of protein structures are suitable for the analysis of normality. Many of these indices are based on interatomic distance and contact analysis. Ramachandran plot analysis of modeled structure can also be used to find the residues that are lying in the favored region, allowed region, and outlier region. The residues corresponding to the outlier region can be taken into consideration for remodeling to improve the accuracy of the model.

3.2 Ligand Compound

A large number of ligand/compound databases have been created, which contains the 2D/3D structural details of many chemical compounds as well as natural products [4]. These compounds can be used as a starting molecule for drug designing or also for screening the compounds based on the structural similarity or pharmacophore-based searching. In modern drug discovery, the potential hits are searched, and then these hits are optimized to improve the affinity, selectivity, efficacy, and oral bioavailability. A drug can enter in a clinical trial only after satisfying the pharmacokinetics and pharmacodynamics property (Table 1).

3.2.1 Database Searching

Drug-like compounds can be searched from different available small chemical compound databases. Different searching criteria such as text, SMILES, substructure, molecular weight, LogP, and other physicochemical properties can be used to retrieve a vast set of compounds for virtual screening. Some tools also provide the facility of pharmacophore-based searching.

Table 1
Description of some chemical compound databases used for drug designing

S. No.	Database name	Description	Web link
1	PubChem	Database of small molecules, nucleotides, carbohydrates, lipids, and peptides. It provides information about 2D and 3D structures, SMILES, physicochemical properties, biological activities, toxicity data, and many others	https://pubchem.ncbi.nlm.nih.gov/
2	ZINC	Database of commercially available compounds with information on molecular weight, calculated LogP, and number of rotatable bonds for virtual screening	https://zinc.docking.org/
3	ChEMBL	Database of molecules with drug-like properties	https://www.ebi.ac.uk/chembl/
4	ChEBI	Collection of molecular entities focused on small chemical compounds	https://www.ebi.ac.uk/chebi/
5	ChemDB	A public database of small molecules and also have prediction tools for solubility, 3D structure, and other molecular properties	http://cdb.ics.uci.edu/
6	BindingDB	A public database of the small, drug-like molecule and its protein target with measured binding properties	https://www.bindingdb.org/bind/index.jsp
7	ChemSpider	A public database of small chemical compounds. It provides the facility to search compounds based on text, specified substructure, and other physicochemical properties	http://www.chemspider.com/
8	SuperNatural II	Database of 325,508 natural compounds with 2D structures, physicochemical properties, and predicted toxicity	http://bioinf-applied.charite.de/supernatural_new/index.php

3.2.2 De Novo Ligand Designing

Several tools such as MedChem Designer, ACD/ChemSketch, and Marvin are available for ligand drawing, editing, import, and export of chemical structure in various file formats [5]. Many tools have the facility to draw the structure of a small chemical compound using its SMILES notation. Some tools such as SmiLib, GLARE, and new lead are available for combinatorial chemistry and analog generation, which uses substructures such as scaffolds, building blocks, and linkers. Combinatorial chemistry helps in designing a large number diverse of chemical entities.

3.3 Cavity/Binding Site Analysis and Validation

The structures of numerous protein complexes are currently obtained through experimental methods and are gathered in several databases. However, many proteins do not have information about their binding sites, thereby limiting drug design speed. It is therefore crucial in drug design to find an automatic binding site prediction method. A large number of predictive methods for predicting and analyzing protein–ligand binding sites are available. These methods incorporate various approaches, providing numerous different types of data ranging from ligand binding site residue lists, ligand binding site 3D atomic coordinates, putative binding ligands lists, EC as well as GO terms. Therefore, scientific efforts have been made to utilize these available methods for the development of several automated tools that can be utilized for the prediction of ligand binding site in the target structure [6, 7].

3.3.1 Cavity Prediction

Due to advances in ligand binding site prediction methods, it becomes easy for biologists to predict the binding site in any protein *via* following step-by-step instructions given by developers for each software. Generally, most of the tools require a 3D structure of the protein and provide amino acid residues present in the binding pocket and their volume. Several online tools such as CASTp and COACH are available on the Internet, where we can upload our structure in PDB format to obtain the information of the binding site. The CASTp is a cavity prediction server that generates two α-shape envelopes in a protein structure, and the volume between these two envelopes measures the size of the cavity [8]. It uses a probe radius of 1.4 Å for cavity prediction. The active site is formed by the composition and 3D arrangement of some specific amino acids of a target protein, which is utilized for the binding of a specific substrate. The predicted cavity in human thioredoxin reductase is shown in Fig. 2; human thioredoxin reductase active sites lie in the cavity region of a protein. A protein may have several cavities, and in most of the cases, the cavity with the largest area and volume is associated with binding.

3.3.2 Approaches and Algorithms

The machine learning models are available for binding site prediction, which uses the structural, sequence, or evolutionary information for developing the model [9]. Binding site prediction methods are based on evolutionary algorithms, energy-based algorithms, and geometric algorithms. The evolutionary algorithm of binding site prediction assumes that binding site residues are conserved in evolutionarily related proteins. Various physiochemical properties such as hydrophobicity, solubility, side-chain pKa, and accessibility to a solvent are used to build a machine learning-based model.

An energy-based method uses the information of interaction potential between the protein and ligand from protein–ligand complexes of PDB. An energy-based method considers the interaction potential between the protein and the ligand. Here, a probe is used to calculate the interaction potential on protein to find a favorable binding region. Q-SiteFinder is an energy-based method of cavity

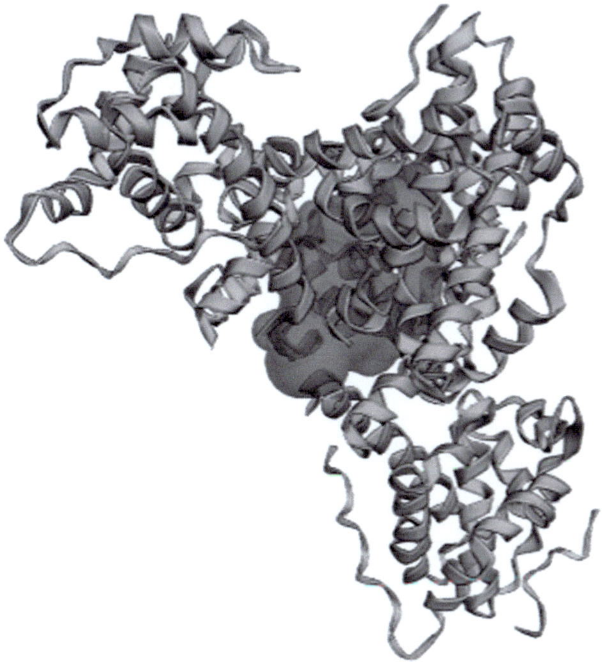

Fig. 2 Predicted cavity (red color) in human thioredoxin reductase (3QFA)

prediction which uses methyl probe to locate the binding site on a protein where a ligand can interact and bind. Geometry based method uses the information such as the shape of the protein, hydrophobic patches on the surface, number of charged residues on the surface, and compactness of a protein for cavity prediction [10].

3.3.3 Validation

The predicted binding site of a protein is validated by comparing it with experimentally verified cavity composition in the functionally and evolutionary related proteins. Binding residues predictions are validated by comparing the predicted residues in the cavity with the experimentally labeled dataset of known cavities from complex structures. Geometry-based methods provide an accurate prediction of the binding site. Geometry based methods can be further improved by incorporating the information of evolutionary conservation of residues in the known binding site of related proteins.

3.4 Molecular Docking

Bioinformatics has recently advanced to the level that allows the prediction of molecular interactions that hold a protein and a ligand together in the bound state to be almost accurate. Many computational tools have been developed to provide a procedure to predict the interaction of small molecules with macromolecular targets (AutoDock, Vina, MVD, etc.) [11]. They can be used to screen a variety of potential compounds, search for new compounds with particular binding characteristics, or test available drugs with functional group modifications.

3.4.1 Protein and Ligand Preparation

The 3D structure of the target protein is retrieved from the PDB database alone or in complex with other molecules. The high-resolution X-ray determined structure is preferred for drug designing. If the target protein exists in the form of complex, then the structural coordinate of the target is separated from the complex using visualizations tools. In case the 3D structure of the target is not available in the PDB database, then proceed for structure modeling using sequence information of target protein. The 3D structure file of the ligand can also be retrieved from chemical compounds databases or can be designed using a structure drawing tool. Molecular docking tools have the facility to prepare the protein and ligand for docking purposes. During protein preparation, common problems such as missing hydrogen atoms, incomplete side chains, and loops, ambiguous protonation states, and flipped residues are corrected, and it depends on the user whether or not to apply any correction. During ligand preparation, different corrections to geometry and filters to ionization states, tautomeric states, chirality, and conformations are applied.

3.4.2 Docking Parameters

Defining the appropriate binding site for docking of a ligand is an important step, which can be known from experiments or predicted by the docking or cavity prediction tool. The docking parameter defines the ligand center and number of torsions, binding region, grid size, flexible residues, docking algorithm, number of runs, and number of poses. Each docking method has its own set of parameters that must be set to proceed for docking.

3.5 Result Interpretation and Interaction Analysis

It is necessary to perform molecular docking and analyze the docked complex for investigating the number of amino acid residues of target and functional groups of ligands involved in different types of noncovalent interaction. Docking results return a number of docked poses for each ligand with their binding energies and affinity. The similarity of docked structures is measured by calculating the root mean square deviation (RMSD) between the coordinates of the atoms. The docking results also provide information about the docked conformation and its position on the target protein. Visualization and interaction wizard of docking tools have the facility to import the protein and docked pose or docked complex to find the different types of interaction between residues of binding sites and atoms of ligand. Docking interaction of aristeromycin with human S-adenosyl-L-homocysteine hydrolase has been shown in Fig. 3. Different types of interactions, participating residues/atoms in a particular interaction, energy of interaction, and ligand pose in the binding region can be traced out.

3.6 Lead Search and Optimization

In lead searching, new bioactive compounds are determined, which may be transformed into a clinically useful drug by lead optimization. A lead compound can be identified through screening of

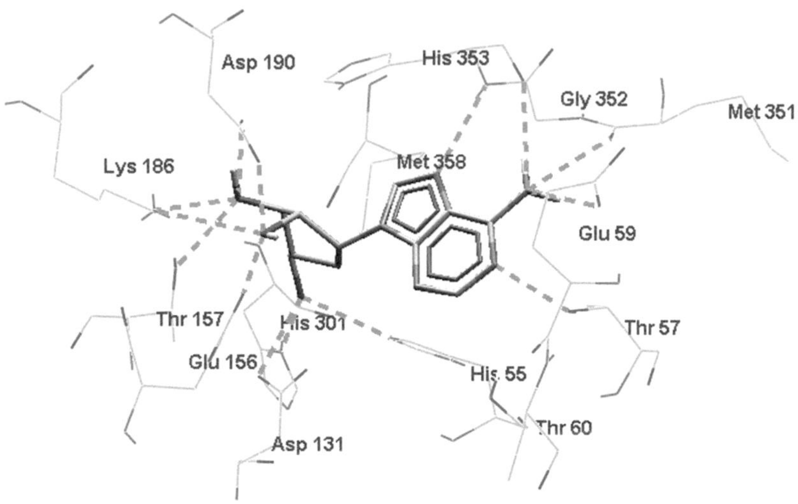

Fig. 3 Docking interaction of aristeromycin with human S-adenosyl-L-homocysteine hydrolase showing amino acid residues involved in hydrogen bonding (green dotted lines)

product metabolites, traditional knowledge from herbal remedies, by chance observation, screening of compound libraries, and optimizing the side effects of a compound. In lead optimization, a lead compound undergoes a series of chemical modifications to improve its physicochemical, pharmacokinetic, and toxicological properties. Here, several analogs of lead compounds are designed by chemical modifications, and their biological and pharmacological response is predicted by QSAR analysis. The process of optimization is performed until a potential drug candidate is searched.

3.7 Lead Compounds

Lead compounds should have the desired potency and selectivity for the target, and it should also be amenable to synthesis and chemical modification. Plants are the best source for lead discovery because they synthesize thousands of different compounds. Many drugs are derived from plant and animal extracts, marine organisms, microbial products. Natural products provide a great diversity of chemical structures. Natural compounds screening can be one of the better approaches for searching a potential lead against a target.

3.7.1 Need for Lead Optimization

A drug should have a very high affinity for a specific target protein, and it should also possess good ADMET properties. Optimization is carried out to improve binding interaction and selectivity of a ligand with its target binding site. Lead Optimization improves the efficacy, pharmacokinetics, and safety profiles of drugs. The structure–property relationship guides the structural modifications required in a lead to improve its pharmacokinetics and pharmacodynamics [12]. Lead optimization should be carried out in such a way so that the optimization of one property should not create an imbalance in another property. For example, if someone is trying to optimize the toxicity by removing a particular group, then this structural change should not reduce the drug solubility or enhance the metabolic stability of the drug. If it happens, then optimize the solubility and or metabolic stability in the next cycle of optimization.

3.7.2 Optimizing the Interaction and Selectivity

The docked protein–ligand complex can guide the optimization of binding interactions between ligand atoms and residues of the target protein. The optimization of binding interaction can improve stability, binding affinity, and drug efficacy. Weak intermolecular interactions, such as hydrophobic interactions, hydrogen bonding, and metal-associated interactions can be taken into consideration for binding optimization. Several strategies that can be used for the optimization of binding interaction and selectivity are given here.

- Vary Alkyl or Aryl Substitution: alkyl group and its length in the lead can be adjusted to improve binding interactions.
- Extension of Extra Functional Group: addition of extra functional group in unused binding space can be utilized to form additional binding interactions.
- Chain Extension or Contraction: additional interactions can be achieved by chain extension and contraction and may also improve the selectivity.
- Ring Expansion or Contraction: ring size variation may bring the functional groups near the binding region and may provide the best fit pose of a ligand in the binding site.
- Ring Variation: Replacement of one ring structure in a ligand with another.
- Isosteres and Bioisosteres: replacement of one functional group by another to optimize the binding interactions without affecting the biological activity.
- Simplification: simple structure fits easily in the binding site, and their synthesis is also easier. For simplification, remove additional functional group or ring structures, and toxic groups.
- Rigidification: introduction of rigid functional groups can limit the free rotations (conformations) in a ligand which may improve the selectivity.

4 Optimizing the Pharmacokinetic Properties

Distribution of a drug to the target site, its metabolism after the action, and the removal of drug metabolites from the body are very important and requisite. Several changes can be performed in the structure of drugs to improve their biological activity. Lipophilicity is one of the important parameters which decides the bioavailability of the drug. A drug with moderate lipophilicity has excellent distribution to tissues and cells. Lipophilicity can be optimized by simple replacement of chemical groups.

- A drug must be metabolically stable to generate the response. It should not undergo metabolism most frequently. The metabolic stability of a drug can be optimized by blocking metabolic sites with the addition of fluorine or other blockers, deactivating the aromatic rings, incorporating a cyclic structure, and changing the ring size.
- Plasma stability of a drug can be improved by substituting an amide for an ester, Increasing steric hindrance near a hydrolysable group.
- Plasma protein binding of a drug can be reduced by decreasing lipophilicity (high lipophilicity high affinity for plasma protein), reducing acidity (more acidic more affinity for albumin).
- Cytochrome p450 enzymes (CYP) play an important role in drug metabolism. CYP inhibition can be reduced by decreasing the lipophilicity of the compound.
- hERG is a gene that codes for a potassium ion channel of heart. hERG channel blocking by a drug can be reduced by replacement of aliphatic hydrogen by methoxy, reducing the basicity of the amine and lipophilicity, and rigidification of linkers.
- Toxicity of a drug can be optimized by removing toxic substructures, avoiding the addition of toxic substructures/groups, the addition of the nonbioactive functional group, blocking the bioactive functional group, and incorporating a bulky substituent near the site of metabolism.

4.1 Pharmacophore Modeling

Pharmacophore is a molecular framework with the essential characteristics of the molecule responsible for the drug's biological activity. The pharmacophore concept has been widely applied to the rational design of new drug molecules. It can be used by schematically depicting the key elements of molecular recognition to represent and identify molecules at a 2D or 3D level. Virtual screening is the most common application of pharmacophores and depending on prior knowledge, different strategies are possible. However, it is also useful for modeling ADME-tox, side effects, and off-target prediction as well as target identification. To improve virtual screening, pharmacophores are often combined with molecular docking simulations [13, 14]. The major steps in pharmacophore modeling are to first take training compounds and assign pharmacophoric features on all the confirmations of each compound. Further, the features of the two most rigid compounds from training sets are aligned to produce intermediate common feature pharmacophore models, followed by model ranking based on selected scoring function [15].

4.2 Pharmacophoric Features

Pharmacophore is a spatial representation of the physicochemical features in molecules required for its biological response. A pharmacophore is shown by the 3D coordinate system with specified features. The pharmacophore mapping is done to find the largest common substructure present within a set of active molecules. Some common pharmacophoric features are hydrogen bond donor (HBD), hydrogen bond acceptor (HBA), aromatic rings (R), hydrophobic groups (H), positively charged groups (P), and negatively charged groups (N). Other pharmacophoric features like bulky groups, chiral centers, metal ions, and solvation penalty areas can also be mapped. These features provide information about the active site of the protein in case of the unavailability of the target structure. These features can help in understanding the amino acid residues in the active site, and their interaction with a molecule. For example, HBA in the pharmacophore specifies HBD in the active site like serine or threonine.

4.3 Searching and Scoring the Common Pharmacophore

Pharmacophoric features are searched in the other active molecules to find a common feature list with similar distance geometry. These common feature sites are searched in all the conformers of the actives, then score and shortlisted based on similar cutoff distance from other features. In PHASE, k-point features are searched in all the actives. In the next step, a set of features that are common in a small subset of active conformations are mapped. Now the K point features that match a minimum subset of actives are determined, and the intersite distance of the features calculated. These pharmacophoric sites for a ligand molecule are called the pharmacophore hypothesis. The pharmacophore hypothesis (DHHHNRR) for inhibitors of thyroid hormone receptors (TRα) was developed, and ten active molecules were aligned with this model to find its compatibility (Fig. 4) [16].

Steps:

- Import and clean the 3D structure of ligands.
- Generate conformers for ligands and minimize the energy.
- Create pharmacophore sites for all the ligands.
- View and edit the pharmacophore sites.
- Select the ligand set for model development and create the site.
- Develop a common pharmacophore hypothesis based on search parameters and scoring function.
- Examine and analyze the Pharmacophore on ligands.

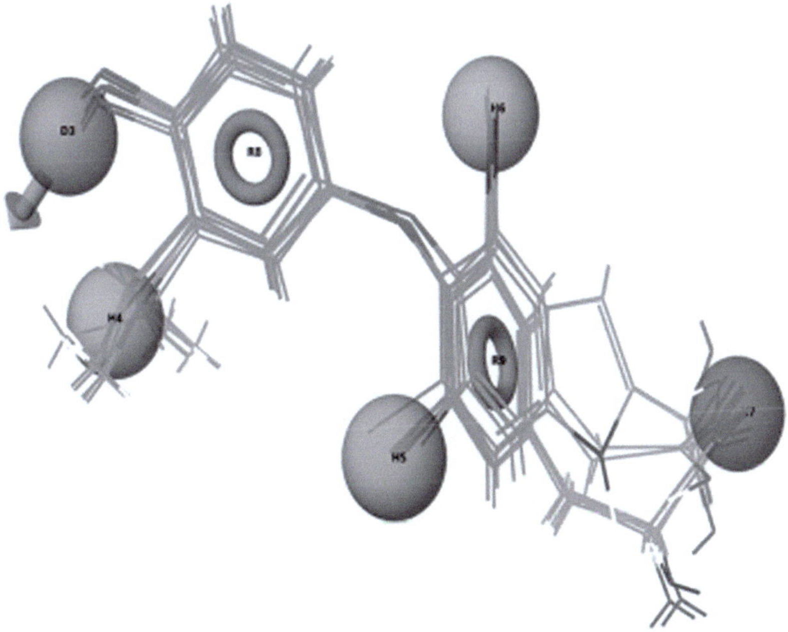

Fig. 4 Alignment of the ten active compounds for the inhibitors of TRα model (Hypothesis: DHHHNRR)

4.4 Application

Pharmacophore help in searching for new novel molecules that possess the same physicochemical features. The pharmacophoric features can be used to screen the 3D database of ligands to find hits that match the pharmacophore hypothesis. Pharmacophoric features can help in designing de novo ligand that may be novel and free from toxic substructure [17]. Different pharmacophore features for an active can be joined by chains, rings, or atom moieties. The pharmacophore acts as a guide to design novel molecules by selecting different groups/rings or atoms with the same feature and joining them by linkers.

5 Molecular Dynamics Simulation

Molecular dynamics simulation (MDS) is one of the key tools for the theoretical and computational study of biomolecules alone or in complex form. MDS can explain the dynamic changes that take place in a protein after a substrate or ligand binds to it. It can also explain the thermodynamic stability of a protein–ligand complex under physiological conditions. It provides information about the atomic fluctuation and dynamic perturbation in macromolecules. The atoms and molecules can interact for a period of time during the simulation. The motion of each atom is calculated and used to predict the overall dynamic behavior of the system [18].

Biological function is based on molecular interactions resulting from macromolecular structures. MDS has evolved into a mature technology that can effectively be used to understand relationships between macromolecular structure and function. Information gathered about macromolecule's dynamic properties is rich enough to shift the usual paradigm of structural bioinformatics from the study of a single structure to the analysis of conformation ensembles [19, 20]. The software used for MDS are CHARMM, Amber, Gromacs, and Desmond.

Steps:

- Prepare the protein or protein–ligand complex using the preparation wizard. This provides the options related to assigning bond orders, add hydrogen, metals, disulfides, waters, selection of chain, heteroatom, and force fields.
- Build a solvated system. This provides the facility to define the solvent model, boundary conditions, box size (distance, angles, volume), neutral medium, ions and its concentration, exclusion criteria for ions (cutoff distance), and transmembrane-related settings.
- Running the MDS. Here, we can define the simulation time, energy and trajectory recording time, ensemble class, temperature, and pressure. At this stage, simulation can be initiated to run.
- Setup the display and view the trajectory of MDS. Summary of simulation analysis can be seen, which also provides information about the total energy, potential energy, temperature, pressure, and volume of the system.
- Result interpretation. The dynamics perturbations and stability of the system during simulation can be explained based on RMSD, RMSF, energy, various types of interactions, and protein–ligand contact, and so on.

5.1 Simulation Parameters

The major steps in molecular dynamics are the generation of topology; define box and solvate, the addition of ions, energy minimization, and equilibration and MD production followed trajectories analysis. Simulation Parameters used in MDS are boundary conditions, box size (distance, angles, volume), neutral medium, ions, and its concentration, exclusion criteria for ions (cutoff distance), transmembrane related settings, simulation time, energy and trajectory recording time, ensemble class, temperature, pressure, and so on.

5.2 Result and Interpretation

For any protein, the RMSD for all residues, backbone, side chain, and C-alpha is calculated with respect to the simulation time. Lower RMSD value indicates a stable form of structure. RMSD values less than 2 Å are preferred. A major difference in the RMSD of a native protein and its mutant structure indicates structural differences as well as dynamic perturbations in the mutant. The root means square fluctuation (RMSF) provides the residue-wise

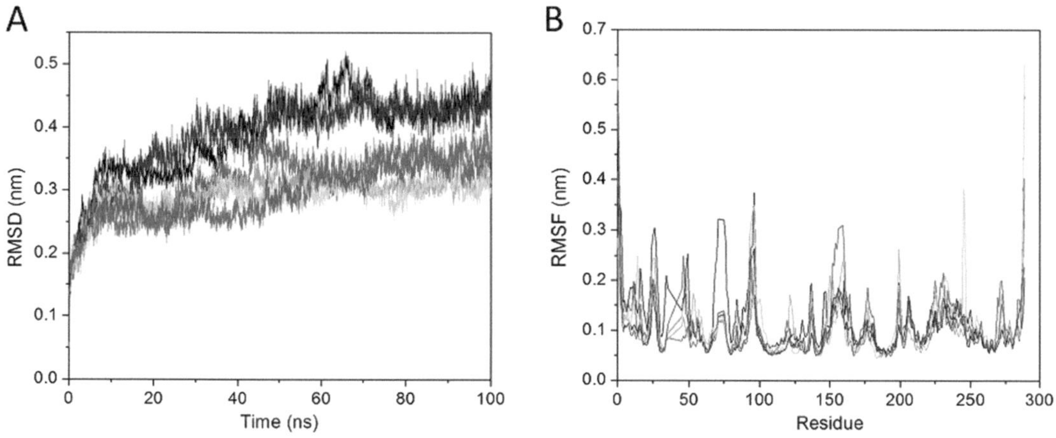

Fig. 5 Molecular dynamics simulation. (A) RMSD of the Cα backbone over the 50 ns MDS at 300 K, (B) RMSF of residues during MD simulation. In all panels the color code is as follows: HsCDK5 (black) and the ligands HsCDK5-Z3R (red), HsCDK5-ZINC85877721 (green), HsCDK5-ZINC96114862 (blue), HsCDK5-ZINC96115616 (cyan), and HsCDK5-ZINC96116231 (magenta)

fluctuation in a protein or complex. It represents the fluctuation of each residue in a protein molecule. The RMSF is plotted as RMSF (nm) vs. residue number which can interpret the region associated with structural alternation or instability. A great extent of residues wise fluctuations may occur in mutant as compared to the native structure. RMSD of the Cα backbone and RMSF of residues human cyclin-dependent kinase 5 (HsCDK5) alone as well in complex with known inhibitor Z3R, ligands ZINC85877721, ZINC96114862, ZINC96115616, and ZINC96116231 were calculated to understand the dynamics of the system (Fig. 5) [21].

During protein folding, the hydrogen bonds play an important role in the proper folding of a structure. In MDS, the residues participating in hydrogen bonding can be calculated with respect to time. Protein–ligand contact map help in understanding the residues of protein that are in contact with ligand molecules for most of the time of interaction. The stability of a protein–ligand complex as well as binding dynamics of ligand on the active site of protein can also be explained based on MDS parameters such as RMSD, RMSF, and protein–ligand contact.

5.3 Application

MDS can find the reason behind the drug resistance of a pathogen and also explains the structural alternations in mutant. MDS is used to understand the unfolding mechanism of a protein. The mechanism of domain unfolding as a result of denaturation or mutation can be explained using MDS. MDS explains the binding interaction, and affinity of a ligand or drug for the binding site of drug target under physiological conditions.

5.4 Pharmacokinetics and Pharmacodynamics Modeling

Pharmacokinetic models provide information about absorption, distribution, metabolism, and excretion processes. ADMET prediction is one of the essential steps in any drug discovery because it provides the pharmacokinetics and pharmacodynamics details of a lead molecule before in vitro and in vivo testing. Pharmacokinetics and pharmacodynamics prediction reduce the experimental cost, time, and risk of drug failure in clinical trials. Pharmacodynamics identifies drug and other properties that regulate the biological responses, that is, therapeutic effects and dynamics of adverse drug reactions to drugs. The accuracy of the Pharmacokinetics model can be improved by including more recent, accurate data set and parameters from experimental studies.

More than 60% of drugs fail in the clinical trial because of poor ADMET properties. Prior prediction of ADMET properties can be done using various software which is based on Kohonen self-organizing maps, artificial neural network, support vector machine, Kernel partial least squares, and multiple linear regression. A predictive model is to build extracting descriptors from the experimental dataset and then used for ADMET prediction of new molecules. ADMET Predictor® tool accurately predicts over 140 ADMET properties. ADMET Predictor contains over 300 atomic and molecular descriptors. In silico methods are more frequently used for ADMET prediction due to the high cost of in vivo and in vitro approaches of ADMET estimation.

5.5 Absorption

There are many barriers in the way of an oral drug to reach its target site. Absorption is one the important property which needs to be determined. Absorption of the drug depends on lipophilicity. A drug should have an optimal range of lipophilicity. Compounds with a lower or higher value of lipophilicity have poor absorption. Lipophilicity of a drug can be optimized to achieve better absorption. Human oral absorption, human intestinal absorption, lipophilicity, and other related parameters can be predicted for any chemical compound [22].

5.6 Distribution

A drug should be distributed properly through systemic circulation to reach its target, but there are many barriers (e.g., blood–brain barrier [BBB]) that need to be overcome. A drug with higher lipophilicity binds with the plasma proteins and other lipophilic molecules. ADMET tools provide the facility to predict BBB, Caco permeability, and P-glycoprotein inhibition [23]. Plasma protein binding prediction models have also been developed using artificial intelligence.

5.7 Metabolism

A drug should be metabolized by drug-metabolizing enzymes after its action. Metabolism of the drug is necessary to avoid the risk of toxicity. A drug that undergoes the most frequent metabolism will not be able to generate the desired therapeutic effect. Metabolically labile drugs can be optimized by removing or substituting the labile group with a stable group. Cytochrome P450 enzymes (CYPs) play a very important role in the phase I metabolism of drugs. CYPs help in understanding the drug–drug interaction. The Metabolism module of ADMET allows us to predict the drug-metabolizing enzyme, its parameters, and CYPs inhibition.

5.8 Excretion

Drugs or their metabolites should be excreted from the body, otherwise, they will exert some toxic effect. The information about the specific metabolite produced as a result of drug metabolism is very important to understand drug toxicity, efficacy, and clearance from the body.

5.9 Toxicity

A drug should not possess any toxic effect. The toxicity of drugs may be due to its interaction with other biomolecules of the body, or its product of metabolism may be toxic. Toxicity of drug to liver, kidney, lungs, and heart can be predicted using different predictor parameters. The toxicity module predicts the maximum recommended therapeutic dose, skin sensitivity, respiratory sensitivity, hERG binding, phospholipidosis, genotoxicity, mutagenicity, and carcinogenicity.

6 Quantitative Structure–Activity Relationship (QSAR)

QSAR analysis is used to predict the biological activity of a molecule using physicochemical properties. QSAR explores the dependency of chemical and biological activities on their molecular features. It establishes a relationship between the activity of a set of molecules and molecular descriptors. The quality of a QSAR model depends on the quality and amount of data available for modeling. QSAR model can be developed if a good number of drugs with known biological activities against a drug target are available [24]. Chemical compound databases can be used to build QSAR models based on available structural properties of the compounds that correlate with their biological property. The QSAR models should be validated and also should be updated as new and more accurate data becomes available.

Basic steps involved in QSAR modeling are as follows;

1. Chemical structure and data curation by removing wrong structure duplicates, tautomers, and so on.
2. Descriptor generation (electronic, geometrical, constitutional, steric, hydrophobic, lipophilicity, quantum chemical, etc.) for selected compounds.
3. Feature selection to include in model building.

4. Creation of a large number of QSAR models using several machine learning approaches such as multiple linear regression, support vector machine, principal component analysis, neural network, and so on.
5. Model validation with an internal test set (experimental versus predicted activity) to assess its quality.
6. Model selection based on internal validation.
7. Model validation with an external dataset to ensure the accuracy of predictability.

6.1 Parameters

For QSAR modeling, molecular structures and corresponding biological activity data of a large number of compounds for a specified target are collected from chemical databases. Data curation involves the removal of irrelevant data (nonbioactivity type), filtering out missing data, and handling of duplicates and tautomers [25]. Molecular descriptors can approximate the most structural properties of a compound. A large number of chemical descriptors related to electronic, steric, hydrophobicity, symmetry, size, shape, and so on are available. It is a very complex task to select the most relevant and adequate features for a problem.

6.2 Models and Approaches

Computational models have been developed for predicting the toxicity of the liver, kidney, lungs, and heart. A drug may have multiple targets in the body and can exert other adverse effects along with the therapeutic response. A sufficient number of features should be selected for building a model using an optimized approach for feature selection. Several approaches such as correlation matrix, principal component analysis, partial least squares, and regression coefficients methods are used for the purpose of feature selection.

6.3 Analysis and Application

There are many tools available for QSAR modeling such as AutoQSAR, OCHEM (online), and eTOXlab (open source). OCHEM builds the QSAR model based on the experimental dataset available in its online chemical database. QSAR-based tools are also used to screen compound databases in order to determine the desired biological activity of chemical molecules based on their structure. Pharmaceutical companies are trying to build a QSAR model for nonhomogeneous data set, which can be applied for a wide range of chemical structures. A scatter plot between predicted activity and observed activity (PIC_{50}) for the inhibitors of thyroid hormone receptor (TRβ) is shown in Fig. 6 [16].

7 Limitations and Future Perspective

CADD has sped up the process of drug discovery and also reduces the cost. Many significant improvements have been made in the

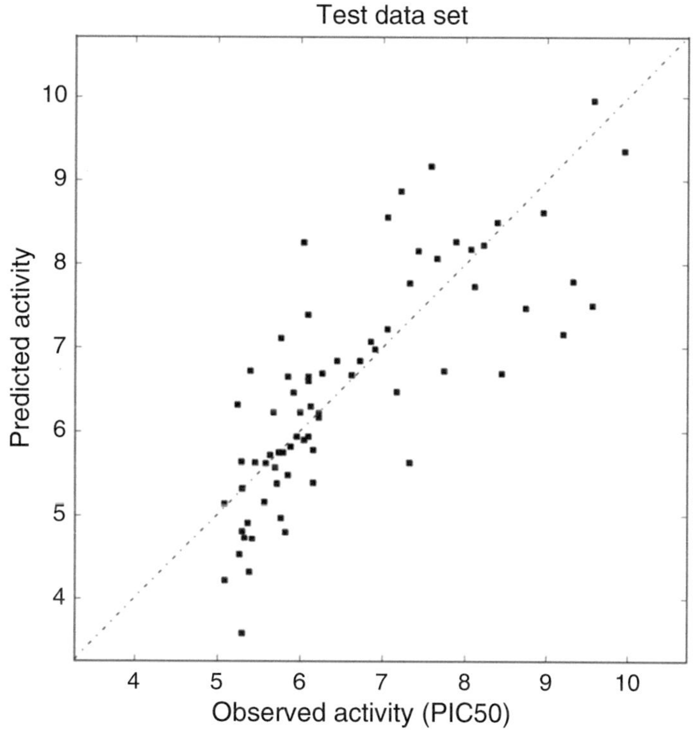

Fig. 6 Scatter plot between predicted activity and observed activity (PIC_{50}) for the inhibitors of TRβ

field of drug discovery software, but still, some issues need to be improved in the future. We have made a significant improvement in the field of computational power, data processing, and data analysis. There is a need to readdress some issues related to parameters used for the prediction and accuracy of tools. Docking and MDS tools have played an important role in understanding the molecular structure and interactions. There exists a possibility to consider some other relevant parameters and criteria while developing docking and MDS tools. In silico ADMET tools predicts the pharmacokinetics of a drug very quickly, but there is a need to include the more experimental datasets in the model development to improve the accuracy of prediction. The predictive accuracy of the QSAR model depends on the selection of features used for developing model, therefore, the process of feature selection and model generation can be more optimized using the more precise algorithm or hybrid approaches. Overall, the accuracy of different predictive approaches and tools used in CADD should be enhanced to speed up the process.

8 Notes

Computational approaches have to speed up the in silico study of different drug targets in the human and pathogen. The availability of the vast amount of structural diversity in compound databases also facilitates the searching of novel chemical entities for drug discovery for a particular target. Various computational tools are available for the binding site/cavity analysis, structural modeling, docking, structure comparison and visualization, MDS, pharmacophore mapping, lead searching and optimization, ADMET and QSAR which allow us to choose an appropriate tool for a case and can compare the results of two or more tools. In recent years, the predictive accuracy of many tools has been significantly improved by including new algorithms, parameters, and more datasets for training and validation purposes.

References

1. Sliwoski G, Kothiwale S, Meiler J, Lowe EW Jr (2013) Computational methods in drug discovery. Pharmacol Rev 66(1):334–395
2. Yu W, MacKerell AD Jr (2017) Computer-aided drug design methods. Methods Mol Biol 1520:85–106
3. Sippl MJ (1993) Recognition of errors in three-dimensional structures of proteins. Proteins 17:355–362
4. Hersey A, Senger S, Overington JP (2012) Open data for drug discovery: learning from the biological community. Future Med Chem 4 (15):1865–1867
5. Singla D, Dhanda SK, Chauhan JS, Bhardwaj A, Brahmachari SK, Open Source Drug Discovery Consortium, Raghava GP (2013) Open source software and web services for designing therapeutic molecules. Curr Top Med Chem 13(10):1172–1191
6. Roche D, Brackenridge D, McGuffin L (2015) Proteins and their interacting partners: an introduction to protein–ligand binding site prediction methods. Int J Mol Sci 16 (12):29829–29842
7. Jiang M, Li Z, Bian Y, Wei Z (2019) A novel protein descriptor for the prediction of drug binding sites. BMC Bioinformatics 20(1):1–13
8. Jie L, Clare W, Herbert E (1998) Anatomy of protein pockets and cavities: measurement of binding site geometry and implications for ligand design. Protein Sci 7(9):1884–1897
9. Liang S, Zhang C, Liu S, Zhou Y (2006) Protein binding site prediction using an empirical scoring function. Nucleic Acids Res 34 (13):3698–3707
10. Gabb HA, Jackson RM, Sternberg MJ (1997) Modeling protein docking using shape complementarity, electrostatics and biochemical information. J Mol Biol 272(1):106–120
11. Rizvi SM, Shakil S, Haneef M (2013) A simple click by click protocol to perform docking: AutoDock 4.2 made easy for non-bioinformaticians. EXCLI J 12:831–857
12. Singh DB (2018) Natural lead compounds and strategies for optimization. In: Ul-Haq Z, Wilson AK (eds) Frontiers in computational chemistry, vol 4. Bentham Science, Sharjah, pp 1–47
13. Yang SY (2010) Pharmacophore modeling and applications in drug discovery: challenges and recent advances. Drug Discov Today 15 (11–12):444–450
14. Qing X, Lee XY, De Raeymaecker J, Tame JR, Zhang KY, De Maeyer M, Voet A (2014) Pharmacophore modeling: advances, limitations, and current utility in drug discovery. J Receptor Ligand Channel Res 7:81–92
15. Akram M, Waratchareeyakul W, Haupenthal J, Hartmann RW, Schuster D (2017) Pharmacophore modeling and in silico/in vitro screening for human cytochrome P450 11B1 and cytochrome P450 11B2 inhibitors. Front Chem 5:104
16. Gupta MK, Misra K (2014) Atom-based 3D-QSAR, molecular docking and molecular dynamics simulation assessment of inhibitors for thyroid hormone receptor α and β. J Mol Model 20(6):2286

17. Warszycki D, Mordalski S, Kristiansen K, Kafel R, Sylte I, Chilmonczyk Z, Bojarski AJ (2013) A linear combination of pharmacophore hypotheses as a new tool in search of new active compounds—an application for 5-Ht1a receptor ligands. PLoS One 8:E84510
18. Huang HJ, Yu HW, Chen CY, Hsu CH, Chen HY, Lee KJ, Tsai FJ, Chen CYC (2010) Current developments of computer-aided drug design. J Taiwan Inst Chem Eng 41 (6):623–635
19. Singh DB, Dwivedi S (2016) Docking and molecular dynamics simulation study of inhibitor 2-Fluoroaristeromycin with anti-malarial drug target PfSAHH. Netw Model Anal Health Inform Bioinforma 5:16
20. Pathak RK, Gupta A, Shukla R, Baunthiyal M (2018) Identification of new drug-like compounds from millets as xanthine oxidoreductase inhibitors for treatment of hyperuricemia: a molecular docking and simulation study. Comput Biol Chem 76:32–41
21. Shukla R, Singh TR (2019) Virtual screening, pharmacokinetics, molecular dynamics and binding free energy analysis for small natural molecules against cyclindependent kinase 5 for Alzheimer's disease. J Biomol Struct Dyn 38:248. https://doi.org/10.1080/07391102.2019.1571947
22. Singh DB, Dwivedi S (2019) Computational screening and ADMET-based study for targeting Plasmodium S-adenosyl-l-homocysteine hydrolase: top scoring inhibitors. Netw Model Anal Health Inform Bioinforma 8:4
23. Singh DB, Dwivedi S (2016) Structural insight into binding mode of inhibitor with SAHH of Plasmodium and human: interaction of curcumin with anti-malarial drug targets. J Chem Biol 9(4):107–120
24. Buglak AA, Zherdev AV, Lei H-T, Dzantiev BB (2019) QSAR analysis of immune recognition for triazine herbicides based on immunoassay data for polyclonal and monoclonal antibodies. PLoS One 14(4):e0214879
25. Kausar S, Falcao AO (2018) An automated framework for QSAR model building. J Chem inform 10(1):1

Chapter 7

Bacteriophage Control for *Pseudomonas aeruginosa* Biofilm Formation and Eradication

Pramila Devi Umrao, Vineet Kumar, Sadhana Singh Sagar, and Shilpa Deshpande Kaistha

Abstract

Microbial biofilms are a type of cell growth observed on biotic or abiotic substrates, with altered gene expression profile from its free-living counterparts, encased in an exopolymeric substance fortification that protects its residents from different stressors. It is now estimated that in almost all niches, microorganisms prefer to exist in a biofilm mode than as free-living cells. The study of biofilms for understanding microbial nature has hence become one of the mainstay assays in a microbiology laboratory. Biofilms show great diversity in their shape, structure, morphology, metabolic physiology as well as genetic complexity based on the resident flora and form the basis of microbial resistance to biocontrol agents such as antimicrobial compounds. Bacteriophages are emerging as effective and safe antibacterial alternative strategies in an era of rapidly emerging antibiotic resistance. Basic biofilm assay protocols that can be performed in any microbiology laboratory are reported along with the use of bacteriophages as biocontrol agents.

Keywords Bacteriophage, Biocontrol, Biofilm formation, Biofilm eradication

1 Introduction

1.1 Microbial Biofilms

The concept of biofilms was first put forth by Bill Costernon wherein he described biofilms as follows: "biofilm consists of single cells and microcolonies, all embedded in a highly hydrated, predominantly anionic exopolymer matrix" [1]. In 2002, the term biofilm was further refined as "immobile communities of organisms (sessile cells) attached to a substratum or to each other, embedded in a matrix of extracellular polymeric substances and showing an altered phenotype in comparison with that of their planktonic (free cells) counterparts" [2]. Advancements in imaging and molecular techniques have revealed microbial biofilms to be dynamic complex microcommunities communicating via coordinated gene regulation to develop an organized interior architectural structure surrounded by exopolymeric substance for its protection from

Neelima Gupta and Varsha Gupta (eds.), *Experimental Protocols in Biotechnology*, Springer Protocols Handbooks,
https://doi.org/10.1007/978-1-0716-0607-0_7,

environmental stressors [3–5]. Biofilm formation can be briefly described in the following steps:

1.1.1 Adsorption and Adhesion

An initial reversible weak followed by irreversible interaction occurs between microbial cell surface and substratum leading to cell adhesion [6]. This stage is influenced by surface charge and tension of substrate, environmental factors such as nutrition, pH, oxygen, redox potential, and osmolarity, motility appendages, and charge and hydrophobicity of the microbial cell [7–10]. Upon overcoming electrostatic charges, microbes adsorb onto suitable biotic or abiotic surfaces by downregulating their motility appendages [11].

1.1.2 Microcolony Formation

The first step in two-dimensional cellular aggregation occurs either by additional adsorption of other microbial species on preexisting microbes or by logarithmic growth causing biofilm nucleation known as microcolonies [7] .

1.1.3 Mature Biofilm

A three-dimensional spatial and heterogeneous structure with an exopolymeric matrix (composed of polysaccharides, proteins, and nucleic acids) around the microbial cells with circulatory and excretory channels within the matrix characterizes a mature biofilm [12, 13]. The unique feature of different microbial biofilms is the well-regulated gene expression of cell signaling quorum sensing (QS) molecules [14, 15]. Gram-negative bacteria typically produce acyl homoserine lactones as QS molecules, while gram-positive bacteria such as *Staphylococcus* spp. produce autoinducer signaling molecules [16].

1.1.4 Dispersal

In response to environmental triggers such as nutrient depletion, oxygen potential, toxin stressors, biofilm residents may again undergo a quorum sensing–related alteration in gene expression, leading to expression of active dispersal related events such as matrix degrading enzyme [17]. Active biofilm dispersal may occur in the form of erosion, wherein release of single or cell clusters takes place at regular intervals from the biofilm. Sloughing occurs at the later stage of biofilm formation when dispersal of large portions of cells occurs. The third type of dispersal is known as seedling dispersal or biofilm hollowing when cells within the biofilm are released as a consequence of molecular mechanisms initiated by the cellular residents. Erosion and sloughing may also occur as part of passive biofilm dispersal caused by external factors such as fluid shear, abrasion and predator phagocytosis [17].

Biofilms put forth the idea of microbial intelligence in terms of response to environmental signals, quorum sensing, decision-making, adaptation, cellular associations, and awareness [18, 19]. Their ability to survive in the face of environmental stressors make them formidable foes in the treatment of infectious diseases. In fact, microbial biofilms have been described to be

100–10,000 fold more resistant to antimicrobial compounds such as antibiotics [20, 21]. Hence alternative strategies for the control of microbial biofilms has become essential, of which one strategy involves the use of ecofriendly and safe biological control agents such as viruses or bacteriophages [22–27].

1.2 Bacteriophages

Viruses are noncellular entities comprising a genome (DNA or RNA) and an outer cover of proteinaceous capsid; viruses which only infect bacteria are known as bacteriophages or "bacteria-eaters" [28]. Bacteriophages are intracellular parasites of bacteria capable of two types of infections:

1. Virulent or lytic infections leading to cell death with the release of multitude of progeny.
2. Temperate infections leading to intracellular integration of viral genome for extended generations of the bacterial host known as lysogeny; followed by lytic infections upon viral lytic gene expression activation.

Virus induced lysis of bacterial host cells can be a means of biological control of biofilm forming bacteria known as phage therapy. The use of phages show several advantages as biofilm control agents as described below:-

1. CPhages are nanosized structures capable of penetrating through the biofilm exopolymeric matrix that itself serves as a concentrated niche of hosts to infect and lyse [29].
2. Phages expressing depolymerases are capable of degrading biofilm exopolymeric substances [30–33].
3. Genetically engineered phages can be used as cargo carriers for delivery of biofilm disrupting genes, drugs, antibodies and antimicrobial compounds [27, 34].
4. Once the host population diminishes, phage number also reduces in number by a phenomenon known as autodosing, rendering the therapy safe [35].
5. Phages can be used with other control strategies for synergistic effects. Phage antibiotic synergy is described using subinhibitory concentrations of ciprofloxacin for biocontrol of *Pseudo monas aeruginosa* biofilms [36].

2 Technique

2.1 Biofilm Measurement

There are several techniques described for the measurement of biofilm formation [37]. In the protocol described in this chapter, the static microtiter based crystal violet assay has been used [38, 39]. Biofilm assays are based on measuring adherent biomass and exopolymeric substances (EPS) on the surface of the microtiter plate, which is stained by basic dyes such as crystal violet or safranin. The major benefit in the use of 96-well microtiter plates is that it is a high-throughput assay that allows for several variables to be tested simultaneously during the measurement of static microbial growth. For instance, abiotic substrates of various materials such as cut catheter tubes made of polyvinyl chloride (PVC), silicone, latex, and so on can be tested simultaneously by simply adding to different microtiter wells in replicates with different cultures [40]. Media and reagents can be optimized for minimal use and effective time management. Use of microtiter plate spectrophotometer helps in quantification of a large number of samples, which could include different microorganisms or different treatments, in triplicates in a single assay [39]. We describe microtiter plate biofilm protocol with modifications to increase efficiency and overcome "edge effect." One disadvantage of using microtiter plates is "edge effect." The wells at the edges of the plate have higher evaporation rate, leading to erroneous results and high variability in assays with poor reproducibility. The other problem is with high rate of contamination in the microtiter plates which may be external or between wells.

For this section, we will describe protocols for preparation of culture media and bacterial inoculum (model organism *Pseudomonas aeruginosa*) in addition to setting up biofilm assays in microtiter plates.

2.2 Bacteriophage Therapy

Use of bacteriophage or phage therapy involves the following steps [41]:

1. Isolation of phage (s) that show host specificity to the target bacteria. Such phages may be isolated from the environment or may be acquired from phage banks.
2. Phage purification.
3. Phage propagation and concentration.

2.3 Effect of Phage on Biofilm Formation and Eradication

A rapid and simple microtiter based assay for studying the effect of phage activity on biofilm formation as well as eradication of preformed biofilm is described wherein phage effect on the ability of cells to bind to microtiter plate surface is assessed [42, 43]. The simple crystal violet biofilm staining assay is insufficient in determining cell death in previously formed biofilm for which the use of viable MTT [3-(4,5-dimethylthiazol-2-yl)-2,5-diphenyltetrazolium bromide] dye has been described [44, 45].

3 Materials

3.1 Preparation of Culture Media and Bacterial Inoculum

1. All media are procured from HiMedia, India, and all analytical grade chemicals are procured from Merck, India.
2. Tryptone Soyabean Casein Digest Medium (TSA).

Constituents	Composition
Tryptone	15.0 g
Peptone	5.0 g
NaCl	5.0 g
Agar for solid media	15 g
Distilled water	1 L
pH	7.3
Autoclave at 121 °C for 20 min	

3. *Pseudomonas aeruginosa* ATCC 15442.
4. Inoculation loop, micropipettes, sterile tips.
5. Instrumentation: Incubator Shaker, Incubator, Autoclave.

3.2 Bacteriophage Isolation

1. Tryptone Soyabean Casein Digest Agar Medium (TSA).
2. Phage source: sewage sample, hospital waste, River Ganges water, and so on.
3. *Pseudomonas aeruginosa* ATCC 15442.
4. SM Buffer (for phage suspension).

Constituents	Composition
NaCl	5.8 g/L
$MgSO_4$	1.2 g/L
1 M Tris HCl pH 7.5	50 mL/L
2% gelatin	5 mL
Distilled water	950 mL
Autoclave at 121 °C for 20 min	

5. Tryptone Soyabean Casein Digest Soft Agar Overlay Medium.

Constituents	Composition
Tryptone	15.0 g/L
Peptone	5.0 g/L
NaCl	5.0 g/L
Distilled water	1 L
pH	7.3
Agar (for soft solid media)	8.0 g/L
Autoclave at 121 °C for 20 min	

6. Chloroform AR.
7. Disposable Syringe filter (0.22 μm, 0.45 μm).
8. *Tryptone Soyapeptone liquid culture media*

Constituents	Composition
Tryptone	15.0 g/L
Peptone	5.0 g/L
NaCl	5.0 g/L
Distilled water	1 L
pH	7.3
Autoclave at 121 °C for 20 min	

3.3 Rapid Microtiter Plate for Static Biofilm Formation Assay and Bacteriophage Activity

1. Sterile Tryptone Soyapeptone Casein Digest culture media.
2. Sterile disposable microtiter plates with lids (U bottom),
3. *Pseudomonas aeruginosa* ATCC 15442.
4. Purified phage with known titer.
5. 0.1% Crystal violet
6. *Crystal violet solution for biofilm assay.*

Constituents	Composition
Crystal violet powder	0.1 g
Distilled water	80 mL
Methanol	20 mL

Dissolve crystal violet powder in H_2O and add methanol. Filter solution to remove crystal debris, which might interfere with biofilm assay. Store solution in the dark at room temperature for up to 2 months

7. 1× Phosphate Buffered Saline (PBS)

Constituents	Composition
NaCl	8.0 g/L
KCl	0.2 g/L
Na_2HPO_4	1.44 g/L
KH_2PO_4	0.24 g/L
pH	7.5
Distilled water	1 L
Autoclave at 121 °C for 20 min	

8. 30% glacial acetic acid
9. Methanol.
10. Di Methyl Sulfoxide (DMSO).
11. Micropipettes, sterile tips, Microtiter plate reader.

3.4 Viability Assay for Biofilms

1. Sterile Tryptone Soyapeptone Casein Digest (TS) culture media.
2. Tetrazolium dye MTT [3-(4,5-dimethylthiazol-2-yl)-2,5-diphenyltetrazolium bromide].
3. *Pseudomonas aeruginosa* ATCC 15442.
4. Purified phage with known titer.
5. Sterile disposable microtiter plates (U bottom) with lids.
6. Micropipettes, sterile tips, Microtiter plate reader.

4 Methods

4.1 Preparation of Culture Media and Bacterial Inoculum

The nutrients base that supports the growth of organisms is called culture medium [46]. Media contain organic and inorganic nutrients composed of complex materials that are rich in vitamins and nutrients. Three of the most commonly used components are beef extract, yeast extract, and peptone, but cultivation of particular microorganisms requires selective media. Agar is added for solidification of media. Basically, media are categorized as natural, semi-synthetic, and synthetic for microbiological works. Media can be made differential for the identification of different microorganisms by utilizing metabolic pathways specific for certain organisms and specific pH indicator dyes. For example, addition of mannitol to media will allow for a distinction between red and yellow colored colonies (neutral red dye based) based on the ability of organisms to ferment mannitol and produce acid. Media can also be enriched

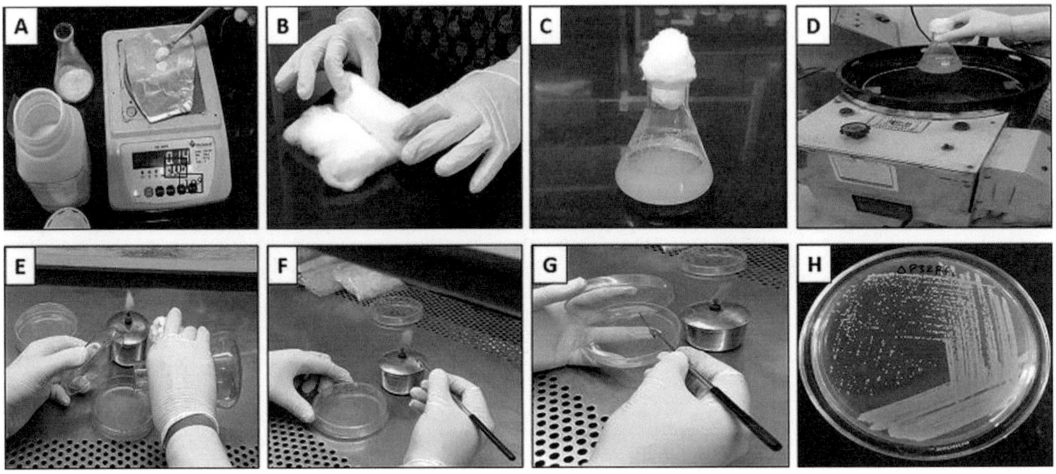

Fig. 1 Stepwise preparation of culture media and pure culture isolation. (**a**) Weighing of media ingredients. (**b**) Preparation of cotton plug. (c) *Media* prior to sterilization in conical flask with cotton plug. (**d**) Sterilization of media using autoclave. (**e**) Pouring of culture media in petri dish inside a laminar air flow cabinet. (**f**) Flaming of inoculation loop. (**g**) Streaking technique for pure culture isolation. (h) *Pseudomonas aeruginosa* streak plate

with specific nutrient requirements such as blood and serum for the growth of fastidious microorganisms. The media for biofilm assay will therefore depend on the type of organisms to be studied.

In this protocol, we describe the use of a complex nutrient media (Tryptone Soyabean Casein Digest Medium—TSA) that has been used for the growth of test organisms *Pseudomonas aeruginosa* ATCC 15442.

Instruments Used: weighing machine, autoclave, hot air oven.

Media and Reagent composition: Subheading 3.1.

Time needed:

Media preparation: 10 min.

Media sterilization: 45 min.

Pouring plates and solidifying time: 20 min.

4.1.1 Procedure (Fig. 1)

1. Weigh components of the media to be prepared and transfer into a clean 500 mL conical flask as described previously (Subheading 3.1).
2. Add 80 mL of distilled water into the measuring cylinder and transfer into the conical flask to dilute the media. Make up volume to 100 mL.
3. Mix media properly by shaking.
4. Close flask mouth using cotton plug. Cotton plug is made by rolling a wad of nonadsorbent cotton.
5. Place the conical flask with the media solution into an autoclave. Sterilize media by autoclaving at 15 lbs pressure and temperature 121 °C for 15 min.

6. Allow the autoclave to cool down before removing media and sterilized glassware.
7. Prepared media is transparent and yellow, ready to use.
8. Inoculate pure culture of host organisms into the liquid culture media using inoculating loop which is sterilized by red hot flaming techniques. Place inoculated media in incubator at 37 °C for 18 h.
9. Observe growth and confirm purity by gram staining using bright field upright microscopy.
10. Streak the pure culture onto solid agar slant/petri dish for obtaining single axenic colonies to be used for short-term preservation.
11. Prepare glycerol stock for long-term preservation: Add 500 μL of the log phage culture to 500 μL of 40% sterile glycerol in a 2 mL screw-top tube or cryo-vial and gently mix to get a final concentration of 20% glycerol. Store at −80 °C. To revive culture, thaw to take a loopful and inoculate in culture media. Refreeze glycerol stock for future use.

4.2 Bacteriophage Isolation (Fig. 2) [36]

Instruments Used: Weighing machine, Autoclave, Hot air oven, sterile disposable syringe filters.

Media and Reagent composition: Subheading 3.2.

Time required:

Prephage preparation: 15 min.

Overnight incubation: 12-16 h.

4.2.1 Spot Assay

Spotting: 5 min.

Drying plates: 15 min.

Overnight incubation: 12–16 h.

Checking for plaques: 5 min.

4.2.2 Double Layer Agar Overlay Assay (DLA)

Serial dilution: 10 min.

Pouring base agar: 10 min.

Mixing phage: host in top agar and pouring: 10 min.

Overnight incubation: 12–16 h.

Checking for plaques: 5 min.

1. Transfer sewage sample/phage source to the laboratory preferably within 1 h for phage isolation: centrifuge sample (10,000 rpm, 10 min, 4 °C) and filter sterilized (0.45 μm pore size Millipore filter) the supernantant.

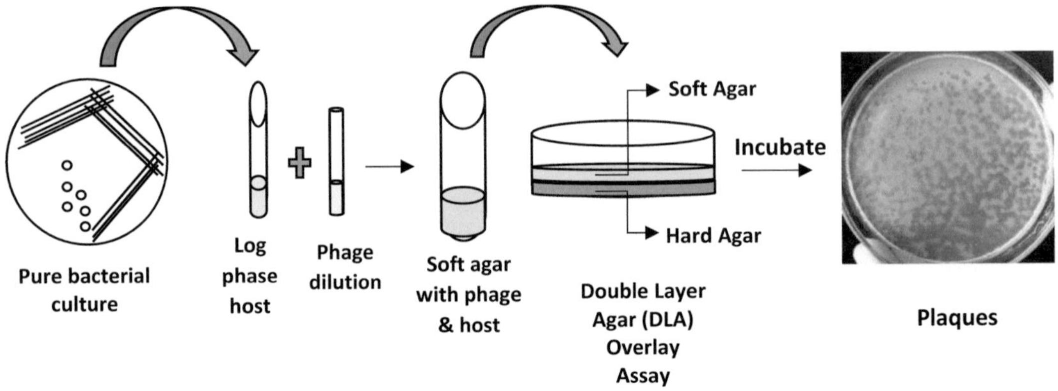

Fig. 2 Double layer agar overlay assay for phage isolation. Log phage host bacterial culture is mixed with phage dilution in soft agar and poured over a previously prepared hard agar base plate. The soft agar is allowed to set prior incubation at suitable temperature creating a double layer agar assembly. After 18–24 h, zone of clearance due to bacterial lysis by phage activity, known as plaques, are observed which are used to further purify phage

2. Mix filtered sewage sample (5 mL) with 5.0 mL overnight host culture and incubate at 37 °C overnight.
3. Next day, add chloroform to cause bacterial cell lysis, centrifuge to remove bacterial cells; filter sterilize supernatant and check for the presence of phages using spot assay or double agar overlay plaque assay.
4. Spot assay: Serially dilute phage suspension in SM buffer. Pour soft TSA (0.8% agar, w/v) premixed with 100 μL of 0.1 OD (10^7 CFU/mL) of log phase *Ps. aeruginosa* host cells (grown overnight at 37 °C, 120 rpm) onto a sterile petri dish and allow to solidify. Spot inoculate 10 μL diluted phage on the surface of molten TSA (0.8% agar, w/v) [36]. Clear zone of plaques will be observed after incubating the plates overnight at 37 °C.
5. Double Layer Agar (DLA) Overlay Plaque assay: The DLA method is modified from the original method described previously [47]. Add 100 μL serially diluted phage sample to 100 μL of log phase *Ps. aeruginosa* bacterial culture (grown overnight at 37 °C, 120 rpm). Add this solution to 5 mL top soft agar (0.8% TS Agar), gently homogenize, and pour into a 90 mm petri dish previously prepared with 15 mL bottom TS agar (hard agar). The plates are gently swirled, dried for 10 min at room temperature, and then inverted and incubated at 37 °C overnight. After up to 24 h incubation, clear zones can be observed on the top agar surface which may be characterized as plaques (Fig. 2).
6. Include the following control plates: (1) Plates with host but no phage suspension: to ensure that phage suspension is free of contamination and to distinguish between true plaques and

uneven agar distribution. (2) Both agar overlay without host and phage to ensure that agar overlay is free of contamination of bacteria as well as virus.

7. Calculate the phage titer in plaque forming units (pfu)/mL using the given formula:

$$\text{pfu/mL} = \frac{\text{Number of plaques} \times \text{reciprocal of dilution}}{\text{volume of phage suspension used for the assay}}$$

Note: Antibiotics and 10% glycerol are known to enhance plaque size and sensitivity of detection [48]. To test the effects of antibiotics on plaque size, the corresponding antibiotic may be added at the concentration desired to the bottom, top or both agar layers after sterilization of the medium. Glycerol can also be added to the top, bottom or both layers before sterilization. Standardized combination of antibiotic and glycerol also enhances plaque size. The prepared phage titer can be further purified by filter sterilizing using disposable 0.22 μm syringe filter.

4.3 Rapid Microtiter Plate for Static Biofilm Formation/Eradication Assay and Bacteriophage Activity (Fig. 3)

Instruments: *Sterile U bottom microtiter plates with lids/untreated tissue culture plates, microplate reader, micropipette (single and multichannel), sterile tips.*

Media and Reagent composition: Subheading 3.3.

Day 1.

Time needed: 1 h

1. Preparation of inoculum: Log phase bacterial suspension, prepared as discussed previously (Subheading 4.1), adjusted to 0.5 McFarland (10^8 cfu/mL).
2. Design of microtiter plate: Take a printout of 96-well microtiter plate and design the experiment on paper by marking rows for test samples and controls. In case more than one microorganism is being used, ensure that one row has been left out containing no media or water to prevent cross-contamination. Leave the peripheral rows to counter edge effect; they will be dispensed with sterile water. Use of a strong biofilm former as positive control and uninoculated control as negative control should be planned on each microtiter plate.
3. Preparation of tenfold inoculum dilution: Dispense 100 μL of media in the wells in vertical direction. Using a 10 μL multichannel micropipette (for pipetting accuracy) perform tenfold dilution of bacterial inoculum (10^8 to 10^3 cfu/mL). In the last dilution well, discard 10 μL of diluent such that each well of the microtiter plate now contains 100 μL inoculum.
4. Preparation of bacteriophage dilution: Prepare tenfold bacteriophage titer dilution in SM buffer (10^9 pfu/mL, 10^8 pfu/mL, 10^7 pfu/mL) separately in microcentrifuge tubes on ice.

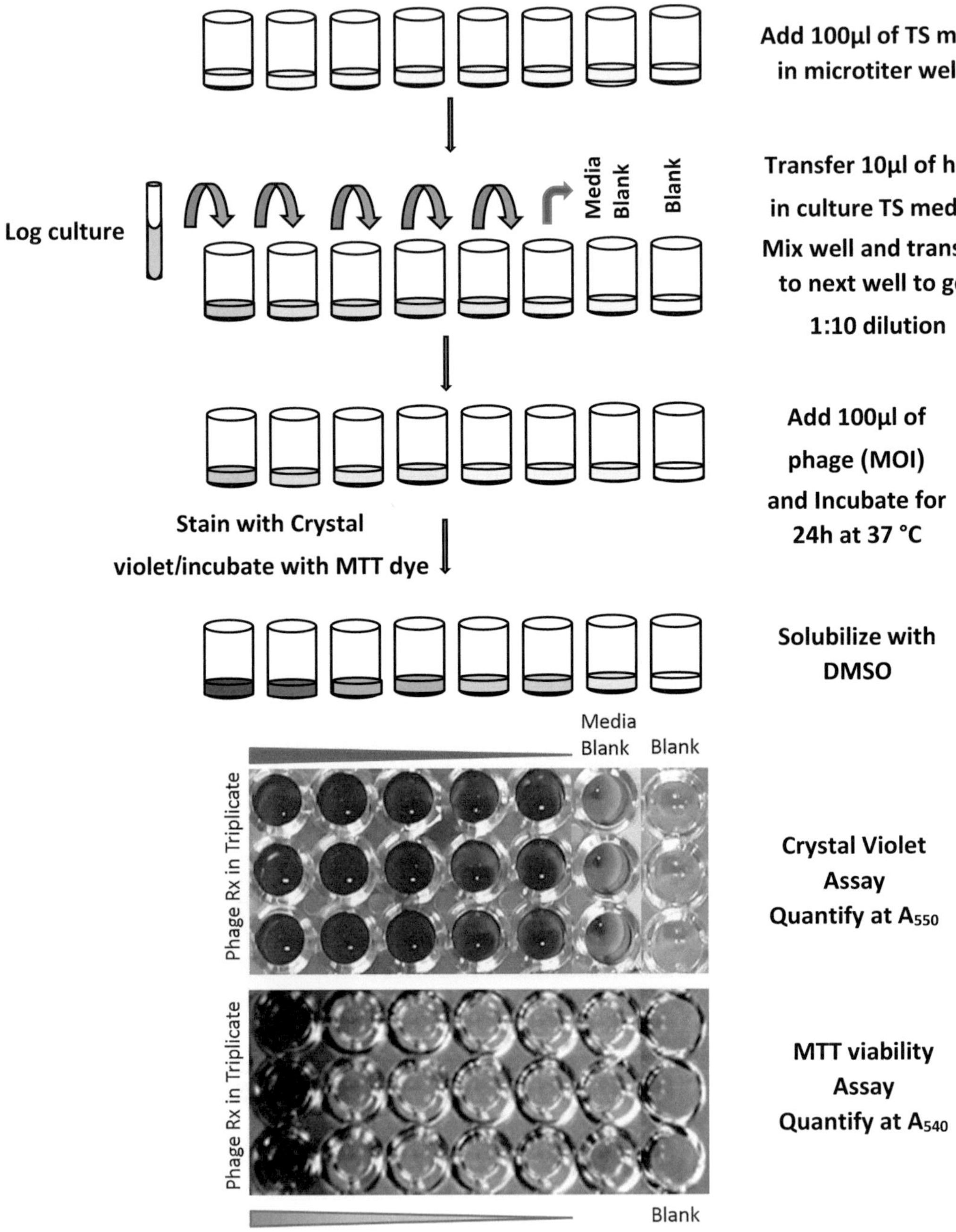

Fig. 3 Schematic for phage—biofilm crystal violet/MTT assay. Flow chart for performing crystal violet assay/MTT viability assay using microtiter plate is shown. Serial dilution of host cells is performed to which different phage preparations and different concentrations of phage (multiplicity of infection) can be added. After incubation for biofilm formation/eradication, number of cells adsorbed onto microtiter plate surface can be measured by staining with crystal violet (CV)/MTT dye. For CV assay, plates are washed to remove excess stain and CV stained biofilm is solubilized with DMSO and quantified at A_{550} using microtiter spectrophotometer. For MTT assay, DMSO is added to wells following incubation with MTT dye and color change recoded at A_{540} using microtiter spectrophotometer

5. Multiplicity of Infection (MOI) for phage treatment of biofilm may be calculated as follows:

$$\text{MOI} = \frac{\text{pfu/mL of phage}}{\text{number of cells(cfu/mL)}}$$

6. Add 100 μL of the diluted bacteriophage suspension to each of the designated wells such that in a single plate we can test different ratio of host and phage combinations (MOI). Cover plates with sterile lids.
7. Stack the microtiter plates in autoclaved tip box (1 mL capacity) with moist cotton/filter paper at the base to prevent evaporation of media in the microtiter plates.
8. Incubate at optimal growth temperature for the desired incubation period (37 °C for 18 h without agitation).
9. For biofilm eradication assay, allow biofilm formation to occur for 18–24 h without bacteriophage addition. Remove carefully with micropipette the suspended media including planktonic cell growth in test wells without disturbing the underlying biofilm. Add 100 μL of bacteriophage suspension to each of the designated wells such that in a single plate we can test different ratios of biofilm and phage combinations (MOI). Incubate as described above.

Day 2.
Time needed:

Planktonic cell transfer: 10 min.

Washing: 15 min.

Fixing biofilm: 30 min.

Crystal violet staining and Washing: 45 min.

Solubilizing and spectrophotometric reading: 15 min.

1. Remove 100 μL planktonic/suspended cells by carefully aspirating with micropipette tip and transfer to a fresh microtiter plate labeled with the same design as test plate. Check to ensure that no cell debris or blobs have been picked up. *Ps. aeruginosa* does not form air-liquid biofilms (ALB) and has a uniform growth pattern. Presence of ALB may be indicative of contamination and so microscopic screening for purity of culture may be carried out for confirmation prior to proceeding with further steps. Measure the planktonic growth optical density by transferring free-living cells in the wells into another fresh microtiter plate and measuring cell growth using spectrophotometer at 620 nm. Ensure use of fresh sterile tips for different wells.

2. In the test biofilm microtiter plate, dispense 1× phosphate buffered saline (PBS) in each well followed by brisk removal of PBS by inverting the plate over water tray/sink. Repeat the washing step three times. Invert the plates and allow the wells to air-dry. (*Note*: Plates can be stored at this stage of the experiment). Ensure that biofilm is not removed due to rough handling during the washing steps.
3. Fix the biomass adherent on the walls of the microtiter plate using 200 μL of methanol for 20 min or drying at 60 °C for 1 h (This step is optional).
4. Add 200 μL of 0.1% w/v crystal violet (same volume as added in the test plate) and allow staining for 30 min at room temperature. (*Note*: Safranin may also be used and dye may be optimized as per organism being used for biofilm assay).
5. Remove the crystal violet solution from the microtiter plates by briskly inverting plate over water tray/sink. Wash dishes with 1× PBS as previously described while shaking out as much liquid as possible after each wash. It is critical to ensure that the biofilm is not removed in the washing step so jerky movements should be avoided. At this stage, the plates may be stored for up to a week at 4 °C as the staining is stable.
6. Add 200 μL of solubilizing agents such as 30% acetic acid or 100% dimethyl sulfoxide (DMSO) or ethanol to each stained well as a dye solubilizing agent [39]. Allow the dye to solubilize by covering plates and incubating 10–15 min at room temperature. (*Note*: it is necessary to optimize the solubilizing agent based on organism used for biofilm assay).
7. Measure dye accumulated in biofilm spectrophotometrically using microtiter plate at 550 nm for crystal violet staining preferably within 1 h. Uninoculated wells containing media alone without staining are used as negative controls.
8. Biofilm formed = (OD of test well − OD of negative control well) ± standard deviation.
9. To interpret biofilm OD results, the following criteria can be used:

$OD_{test} \leq OD_{negative\ control}$: Non-biofilm former.

$OD_{test} > OD_{negative\ control}$ ($p \leq 0.1$) or $OD_{test} < OD_{positive\ control}$ ($p \leq 0.01$): Weak biofilm production.

$OD_{test} > OD_{negative\ control}$ ($p \leq 0.05$) or $OD_{test} < OD_{positive\ control}$ ($p \leq 0.05$): Moderate biofilm production.

$OD_{test} > OD_{negative\ control}$ ($p \leq 0.01$) or $OD_{test} \geq OD_{positive\ control}$ ($p \leq 0.1$): Strong biofilm production.

10. The ratio of planktonic OD to biofilm OD also gives comparative information regarding the ability of the organisms to bind to substrate or remain as free-living cells. Following phage treatment/antimicrobial treatment, a comparison of cell density in planktonic or biofilm mode can also help interpret if the treatment has inhibited biofilm formation in early stages. In later stages when compared with viability data it can also help interpret if the treatment has resulted in cell death or in dispersal of the biofilm.
11. Percentage reduction in biofilm phage biocontrol is calculated as follows:

$$\% \text{ reduction in biofilm formation with phage biocontrol} = \frac{\text{mean OD biofilm treated} - \text{mean OD biofilm phage treated}}{\text{mean OD biofilm treated}} \times 100$$

$$\% \text{ reduction in biofilm eradication with phage control} = \frac{\text{mean OD biofilm treated} - \text{mean OD biofilm phage treated}}{\text{mean OD biofilm untreated}} \times 100$$

4.4 Viability Assay Using Direct MTT Assay (Fig. 3)

The use of light colored MTT dye (Tetrazolium dye MTT 3-(4,5 dimethylthiazol 2 yl)-2,5-diphenyltetrazolium bromide) in prokaryotic and eukaryotic cell viability assay is based on its reduction by cellular dehydrogenases to insoluble formazan, which has a purple color and can be easily measured spectrophotometrically [44, 45]. Hence the amount of purple color of formazan is directly proportional to viable cells present in the assay well. The advantages of the direct MTT assay is its high sensitivity, speed of performing, simplicity and ability to use in high through put assay such as microtiter plate assay for estimating antimicrobial activity.

Time needed:

Setting up biofilm and incubation: 16-18 h

Washing biofilm: 15 min

MTT staining: 2 h

Solubilizing and spectrophotometric reading: 15 min

1. Establish biofilm formation or eradication with phage treatment or no treatment as described previously. Additional controls for MTT dye will include blank with no cells, only media, and only phage treatment with no cells.
2. After the prescribed incubation time, remove free-living cells and dispense 1X PBS in each well. Briskly remove 1X PBS by

inverting plate over water tray/sink ensuring that the biofilm is not disturbed. Repeat the washing step three times.

3. Add 100 μL of 0.3% MTT dye in each well and incubate at 37 °C for 2 h.
4. Add 100 μL DMSO for formazan solubilization and measure color specrophotometrically at 540 nm. Subtract blank readings from phage-treated and untreated readings.
5. Obtain mean and standard deviations for all reading and calculate percentage reduction of biofilm.

$$\% \text{ reduction in biofilm phage biocontrol} = \frac{\text{mean OD phage untreated} - \text{mean OD phage treated}}{\text{mean OD phage untreated}} \times 100$$

5 Notes

1. Ensure that the weighing balance is properly calibrated.
2. Make a tight cotton plug and test that it is not loose. Cotton plug should be dry followed by wrapping with aluminum foil.
3. Always use a flask that holds twice the volume of media you are preparing.
4. In CV biofilm assay [38], it is most critical to ensure that in the washing step there is no loss of biofilm. Practice the hand flick motion prior to performing the assay to ensure no jerk in the hand motion.
5. Adding of autoclaved water to the edges of the microtiter plate ensures that loss of media due to evaporation does not affect internal wells resulting in consistent data with reduced standard deviations in the assay. Use of moist box also ensures consistency and lower data variability.
6. Phage lytic activity is highly host specific [28]. Ensure the purity of bacterial culture being used as host at regular intervals.
7. In spot assay, spot is indicative of lysis activity and are not plaques. For generating plaques, double layer agar overlay must be carried out [47, 49].
8. In double layer agar overlay assay, the top overlay agar should be of right temperature, which can be checked using skin sensitivity of the palm side lower arm. Mixing of host and phage should not be very vigorous but done gently by tapping or by rolling the tubes between the palm whilst wearing a ring on the finger. Pouring of the smooth overlay must be done to

ensure while ensuring that temperature is higher than required for solidification, otherwise lumps of agar as well air bubbles on the top agar overlay hinder plaque visualization [49].

9. In case plaque size is very small, the following troubleshooting can be performed to enhance plaque size:
   (a) Successively decrease agar concentration in top overlay from 0.8% up to 0.3% to obtain larger plaques [47].
   (b) Add optimized concentrations of antibiotics (based on host antibiotic resistivity pattern), glycerol (10%) sodium azide (0.005%) to enhance plaque size [48].
   (c) Methylene blue or tetrazolium dyes may be added to enhance contrast, as viable cells will be blue or purple while plaques will appear colorless [44].
10. MTT dye is sensitive to light and affected by the presence of media constituents. Appropriate controls may be kept to ascertain and blank background readings [44, 50].

Acknowledgments

The authors gratefully acknowledge financial support from University Grants Commission, New Delhi 357996 (VK) and CSJMU Minor Research Project Grant CSJMU/CDC/143/2018 (SDK).

References

1. Lappin-Scott H, Burton S, Stoodley P (2014) Revealing a world of biofilms—the pioneering research of Bill Costerton. Nat Rev Microbiol 12:781–787. https://doi.org/10.1038/nrmicro3343
2. Donlan RM, Costerton JW (2002) Biofilms: survival mechanisms of clinically relevant microorganisms. Clin Microbiol Rev 15:167–193. https://doi.org/10.1128/cmr.15.2.167-193.2002
3. Schlafer S, Meyer RL (2016) Confocal microscopy imaging of the biofilm matrix. J Microbiol Methods 38:50–59. https://doi.org/10.1016/j.mimet.2016.03.002
4. James SA, Powell LC, Wright CJ (2016) Atomic force microscopy of biofilms-imaging, interactions, and mechanics. In: Dhanasekaran D (ed) Microbial biofilms—importance and applications. Intech Open, London. https://doi.org/10.57772/63312
5. Yang L, Hu Y, Liu Y et al (2011) Distinct roles of extracellular polymeric substances in *Pseudomonas aeruginosa* biofilm development. Environ Microbiol 13:1705–1717. https://doi.org/10.1111/j.1462-2920.2011.02503.x
6. Carniello V, Peterson BW, van der Mei HC, Busscher HJ (2018) Physico-chemistry from initial bacterial adhesion to surface-programmed biofilm growth. Adv Colloid Interf Sci 261:1–14. https://doi.org/10.1016/J.CIS.2018.10.005
7. Wei Q, Ma LZ (2013) Biofilm matrix and its regulation in *Pseudomonas aeruginosa*. Int J Mol Sci 14:20983–21005. https://doi.org/10.3390/ijms141020983
8. Garrett TR, Bhakoo M, Zhang Z (2008) Bacterial adhesion and biofilms on surfaces. Prog Nat Sci 18:1049–1056. https://doi.org/10.1016/j.pnsc.2008.04.001
9. Berne C, Ducret A, Hardy GG, Brun YV (2015) Adhesins involved in attachment to abiotic surfaces by gram-negative bacteria. Microbiol Spectr 3:1–46. https://doi.org/10.1128/microbiolspec.MB-0018-2015

10. Arciola CR, Campoccia D, Ravaioli S, Montanaro L (2015) Polysaccharide intercellular adhesin in biofilm: structural and regulatory aspects. Front Cell Infect Microbiol 5:7. https://doi.org/10.3389/fcimb.2015.00007
11. Acemel RD, Govantes F, Cuetos A (2018) Computer simulation study of early bacterial biofilm development. Sci Rep 8:5340. https://doi.org/10.1038/s41598-018-23524-x
12. Flemming H-C, Wingender J (2010) The biofilm matrix. Nat Rev Microbiol 8(9):623–633. https://doi.org/10.1038/nrmicro2415
13. Wimpenny J, Manz W, Szewzyk U (2000) Heterogeneity in biofilms. FEMS Microbiol Rev 24:661–671. https://doi.org/10.1111/j.1574-6976.2000.tb00565.x
14. Quiñones B, Dulla G, Lindow SE (2005) Quorum sensing regulates exopolysaccharide production, motility, and virulence in pseudomonas syringae. Mol Plant-Microbe Interact 18:682–693. https://doi.org/10.1094/MPMI-18-0682
15. Rutherford ST, Bassler BL (2012) Bacterial quorum sensing: its role in virulence and possibilities for its control. Cold Spring Harb Perspect Med 2:11. https://doi.org/10.1101/cshperspect.a012427
16. Papenfort K, Bassler BL (2016) Quorum sensing signal–response systems in Gram-negative bacteria. Nat Rev Microbiol 14:576–588. https://doi.org/10.1038/nrmicro.2016.89
17. Kaplan JB (2010) Biofilm dispersal: mechanisms, clinical implications, and potential therapeutic uses. J Dent Res 89:205–218. https://doi.org/10.1177/0022034509359403
18. Westerhoff HV, Brooks AN, Simeonidis E et al (2014) Macromolecular networks and intelligence in microorganisms. Front Microbiol 5:379. https://doi.org/10.3389/fmicb.2014.00379
19. Majumdar S, Pal S (2017) Bacterial intelligence: imitation games, time-sharing, and long-range quantum coherence. J Cell Commun Signal 11(3):281–284. https://doi.org/10.1007/s12079-017-0394-6
20. Blair JMA, Webber MA, Baylay AJ et al (2014) Molecular mechanisms of antibiotic resistance. Nat Rev Microbiol 13:42–51. https://doi.org/10.1038/nrmicro3380
21. Wang H, Tu F, Gui Z et al (2013) Antibiotic resistance profiles and quorum sensing-dependent virulence factors in clinical isolates of *Pseudomonas aeruginosa*. Indian J Microbiol 53:163–167. https://doi.org/10.1007/s12088-013-0370-7
22. Abedon ST (2016) Bacteriophage exploitation of bacterial biofilms: phage preference for less mature targets? FEMS Microbiol Lett 363(3). https://doi.org/10.1093/femsle/fnv246
23. Perera MN, Abuladze T, Li M et al (2015) Bacteriophage cocktail significantly reduces or eliminates *Listeria monocytogenes* contamination on lettuce, apples, cheese, smoked salmon and frozen foods. Food Microbiol 52:42–48. https://doi.org/10.1016/j.fm.2015.06.006
24. Nanda AM, Thormann K, Frunzke J (2015) Impact of spontaneous prophage induction on the fitness of bacterial populations and host-microbe interactions. J Bacteriol 197:410–419. https://doi.org/10.1128/JB.02230-14
25. Szafrański SP, Winkel A, Stiesch M (2017) The use of bacteriophages to biocontrol oral biofilms. J Biotechnol 250:29–44. https://doi.org/10.1016/j.jbiotec.2017.01.002
26. Górski A, Międzybrodzki R, Weber-Dąbrowska B et al (2016) Phage therapy: combating infections with potential for evolving from merely a treatment for complications to targeting diseases. Front Microbiol 7:1515. https://doi.org/10.3389/fmicb.2016.01515
27. Bárdy P, Pantůček R, Benešík M, Doškař J (2016) Genetically modified bacteriophages in applied microbiology. J Appl Microbiol 121:618–633. https://doi.org/10.1111/jam.13207
28. Ackermann H-W, Węgrzyn G (2014) General characteristics of bacteriophages. In: Phage therapy: current research and applications. Caister Academic Press, Norfolk, pp 43–56
29. Chan BK, Abedon ST (2015) Bacteriophages and their enzymes in biofilm control. Curr Pharm Des 21(1):85–99. https://doi.org/10.2174/1381612820666140905112311
30. Yan J, Mao J, Xie J (2014) Bacteriophage polysaccharide depolymerases and biomedical applications. BioDrugs 28(3):265–274. https://doi.org/10.1007/s40259-013-0081-y
31. Hughes KA, Sutherland IW, Clark J, Jones MV (1998) Bacteriophage and associated polysaccharide depolymerases—novel tools for study of bacterial biofilms. J Appl Microbiol 85:583–590. https://doi.org/10.1046/j.1365-2672.1998.853541.x
32. Pires DP, Oliveira H, Melo LDR et al (2016) Bacteriophage-encoded depolymerases: their diversity and biotechnological applications. Appl Microbiol Biotechnol 100:2141–2151. https://doi.org/10.1007/s00253-015-7247-0

33. Latka A, Maciejewska B, Majkowska-Skrobek G et al (2017) Bacteriophage-encoded virion-associated enzymes to overcome the carbohydrate barriers during the infection process. Appl Microbiol Biotechnol 101:3103–3119. https://doi.org/10.1007/s00253-017-8224-6
34. Motlagh AM, Bhattacharjee AS, Goel R (2016) Biofilm control with natural and genetically-modified phages. World J Microbiol Biotechnol 32:67. https://doi.org/10.1007/s11274-016-2009-4
35. Loc-Carrillo C, Abedon ST (2011) Pros and cons of phage therapy. Bacteriophage 1:111–114. https://doi.org/10.4161/bact.1.2.14590
36. Sagar SS, Kumar R, Kaistha SD (2017) Efficacy of phage and ciprofloxacin co-therapy on the formation and eradication of *Pseudomonas aeruginosa* biofilms. Arab J Sci Eng 42:95–103. https://doi.org/10.1007/s13369-016-2194-3
37. Azeredo J, Azevedo NF, Briandet R et al (2017) Critical review on biofilm methods. Crit Rev Microbiol 43:313–351. https://doi.org/10.1080/1040841X.2016.1208146
38. O'Toole GA (2011) Microtiter dish biofilm formation assay. J Vis Exp 47:1–2. https://doi.org/10.3791/2437
39. Merritt JH, Kadouri DE, O'Toole GA (2005) Growing and analyzing static biofilms. Curr Protoc Microbiol, Chapter 1, Unit 1B.1. https://doi.org/10.1002/9780471729259.mc01b01s00
40. Singh D, Kaistha SD (2018) Multiple antibiotic resistance and biofilm formation of catheter associated urinary tract infection (CAUTI) causing microorganisms. J Bacteriol Mycol 6. https://doi.org/10.15406/jbmoa.2018.06.00208
41. Gill JJ, Hyman P (2010) Phage choice, isolation, and preparation for phage therapy. Curr Pharm Biotechnol 11:2–14. https://doi.org/10.2174/138920110790725311
42. Knezevic P, Petrovic O (2008) A colorimetric microtiter plate method for assessment of phage effect on *Pseudomonas aeruginosa* biofilm. J Microbiol Methods 74:114–118. https://doi.org/10.1016/j.mimet.2008.03.005
43. Xie Y, Wahab L, Gill JJ (2018) Development and validation of a microtiter plate-based assay for determination of bacteriophage host range and virulence. Viruses 10:189. https://doi.org/10.3390/v10040189
44. Grela E, Kozłowska J, Grabowiecka A (2018) Current methodology of MTT assay in bacteria – a review. Acta Histochem 120:303–311. https://doi.org/10.1016/J.ACTHIS.2018.03.007
45. Hundie GB, Woldemeskel D, Gessesse A (2016) Evaluation of direct colorimetric MTT assay for rapid detection of rifampicin and isoniazid resistance in *Mycobacterium tuberculosis*. PLoS One 11:e0169188. https://doi.org/10.1371/journal.pone.0169188
46. Atlas R (2010) Handbook of microbiological media, 4th edn. CRC Press, Boca Raton, Florida
47. Kropinski AM, Mazzocco A, Waddell TE et al (2009) Enumeration of bacteriophages by double agar overlay plaque assay. Methods Mol Biol 501:69–76. https://doi.org/10.1007/978-1-60327-164-6_7
48. Santos SB, Carvalho CM, Sillankorva S et al (2009) The use of antibiotics to improve phage detection and enumeration by the double-layer agar technique. BMC Microbiol 9:148. https://doi.org/10.1186/1471-2180-9-148
49. Kauffman KM, Polz MF (2018) Streamlining standard bacteriophage methods for higher throughput. Methods 5:159–172. https://doi.org/10.1016/J.MEX.2018.01.007
50. Benov L (2019) Effect of growth media on the MTT colorimetric assay in bacteria. PLoS One 14:e0219713. https://doi.org/10.1371/journal.pone.0219713

Chapter 8

Detection of Phosphoproteins (Phosphoserine and Phosphothreonine) from Thylakoid Membranes Using Western Blotting

Varsha Gupta and Baishnab Charan Tripathy

Abstract

In the plant chloroplast, high-energy reactions occur which are responsible for driving the light reaction of photosynthesis for generation of reducing power which help to bring about the dark reaction of photosynthesis for production of carbohydrates. The photosystem components of PS-II along with light harvesting chlorophyll (LHCP) complex undergo reversible phosphorylation and dephosphorylation of several proteins during light and dark cycles of the plant. Many photosystem-II (PS-II) proteins such as D1, D2 reaction center proteins, Chl a-binding proteins (Chl-BP-CP-43), *psbH* gene product, and other chlorophyll a/b binding proteins as LHCB1 and LHCB2 (these are products of *Lhcb1* and *Lhcb2* genes) are phosphorylated at an N-terminal threonine residue. Usage of orthophosphate (^{32}P) for labeling of protein was done for detection of phosphoproteins under in vivo condition. Under in vitro condition, kinase assay can be performed using P^{32} ATP. However, orthophosphate has limitation of missing out proteins, which were previously phosphorylated and is associated with hazards of radioisotopes, while kinase assay is done under in vitro condition where the experimental environment is quite different from the natural in vivo environment. Western blotting and immune-probing by anti-phosphoantibodies are powerful and safer technologies for analyzing phosphorylation status of the proteins in the samples. In this chapter, readers would learn about thylakoid membrane isolation, Bradford reagent preparation, protein estimation, equalizing protein concentration, sample preparation for SDS-PAGE, SDS-PAGE, western blotting, and immune-probing using anti-phosphothreonine and anti-phosphoserine antibodies. These protocols are standardized and are easy to perform in the laboratory.

Keywords Phosphoproteins, Phosphoserine, Phosphothreonine, SDS-PAGE, Western blotting, Antibody

1 Introduction

Plants are exposed to extreme environmental conditions with essential dependence on light. Light is mandatory for photosynthesis in which many photochemical reactions occur in plants. However, light is not constant at every part of earth, and there are extreme fluctuations in light signals which essentially require plants

Neelima Gupta and Varsha Gupta (eds.), *Experimental Protocols in Biotechnology*, Springer Protocols Handbooks,
https://doi.org/10.1007/978-1-0716-0607-0_8,

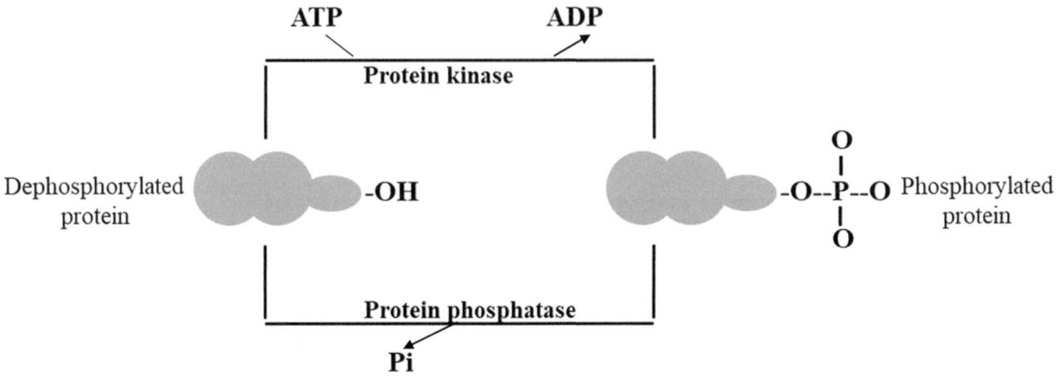

Fig. 1 Reversible phosphorylation reactions. In the polypeptide chain or protein, selected threonine, serine and/or tyrosine residues can accept phosphate group and become phosphorylated. The removal of phosphate group is done for dephosphorylation of proteins. Former reaction is catalyzed by protein kinase and ATP, while latter reaction is catalyzed by protein phosphatase

to acclimatize photosynthesis in changing environmental conditions.

Cellular activities are often regulated by reversible phosphorylation and dephosphorylation of the proteins (Fig. 1). Almost all the plant signaling cascades are controlled by protein phosphorylation [1, 2]. The attachment of phosphate group (phosphorylation) or removal of a phosphate group (dephosphorylation) from a protein influences its structural and functional properties as protein–protein interactions [3], enzymatic activity [4], its localization and stability [5, 6]. Phosphorylation/dephosphorylation is able to regulate the interaction of transcriptional factors with promoter elements [7] controlling gene expression in response to specific signals.

Light, being the critical environmental signal, controls the growth and development of the plant [8]. Sophisticated photosensory systems are evolved by plants of different habitats in response to intensity, wavelength, direction, and duration of light [9, 10]. The three important classes of plant signal-transducing photoreceptors: red/far-red light absorbing phytochromes [11–15], blue light absorbing cryptochromes [16, 17], and phototropins [18, 19] control phosphorylation of many cytoplasmic and nuclear proteins [20].

Plant chloroplast which contains green pigments is an efficient light absorber and these pigments act as antenna molecules (Fig. 2). Chloroplast thylakoid membranes are composed of two regions with stacked thylakoids in grana and unstacked thylakoids in stroma (Fig. 1). The grana discs are flattened and interconnected by the stroma thylakoids called lamellae [21, 22]. Grana has thylakoid stacks whose membrane has embedded PSII/LHCII complexes. These have more than 30 subunits [23, 24]. Photosystem II

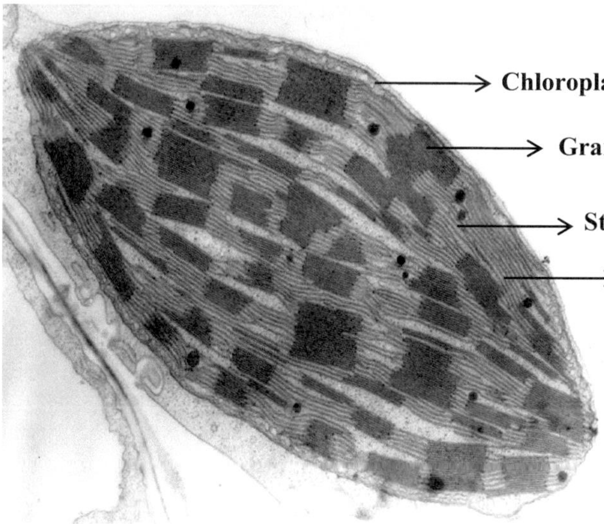

Fig. 2 Electron micrograph of chloroplast, showing stacked grana (grana thylakoids), while unstacked thylakoids are stroma thylakoids. Grana thylakoids have PSII proteins and stroma thylakoids have PSI and ATP synthase

complexes as PSII/LHCII are present in stacked grana, while photosystem I and ATP synthase are present in thylakoids toward stroma or grana end membranes [23]. The flat surfaces of PSII and LHCII complexes allow for their positioning in the stacked grana [25], while complexes of PSI and ATP synthase with their protruded parts are toward the stromal side [26] and few other proteins are dispersed throughout grana and stroma thylakoids. Photosystem II plays a key role in photolysis of water; thus, it is prone for photo-oxidative stress under high intensity of light which leads to photoinhibition [27, 28]. Plants have devised several strategies to overcome these problems.

Protein phosphorylation plays an important role in disassembly of the PSII supercomplex in the grana core under high light [29, 30]. During migration of PSII from the stacked grana to the grana margins and to the stroma thylakoids, they are dephosphorylated by the PSII core phosphatase (PBCP) [31, 32]. These reactions are important for repair cycle and reassembly of PSII complexes. A previous study has shown that many of the thylakoid membrane proteins undergo serine-specific phosphorylations in different light conditions [33]. Thus, it is important to analyze the phosphorylation and dephosphorylation events which control various cellular events. In this chapter readers would learn about in vivo protein phosphorylation technique using antibody probing which can be further used for analysis of any sample in different experimental conditions.

2 Technique

Protein phosphorylation is done by radioisotope labeling using either orthophosphate (^{32}P) or kinase assay using γ-P^{32}-ATP. For in vivo phosphorylation studies orthophosphate is used, but its usage leads to uneven uptake and is associated with hazards of radioisotope. Another issue is sometimes the orthophosphate usage is unable to provide accurate estimate of actual phosphorylation due to the fact that some phosphorylations might be preexisting in the samples. Western blotting and immune-probing by anti-phospho-antibodies are powerful and safer technologies for analyzing phosphorylation status of the proteins in samples. It is therefore very important to analyze the proteins/enzymes which are phosphorylated at serine, threonine, or tyrosine residues.

3 Materials

The following materials would be required for different experiments.

3.1 Thylakoid Membrane Isolation

Frozen tissue, mortar and pestle, ice-cold buffers with protease and phosphatase inhibitors, pipettes, Mira cloth, glass funnel, microcentrifuge tubes, electronic balance, refrigerated centrifuge.

3.2 Protein Estimation

Coomassie brilliant blue G, absolute alcohol, orthophosphoric acid, Bradford reagent, BSA, buffer, protein samples, pipettes and tips, glass cuvettes, UV-visible spectrophotometer.

3.3 SDS-Page

30% acrylamide–bisacrylamide stock solution, buffers, 10% SDS, sample protein, sample buffer, gel electrophoresis unit, power supply, pipettes.

3.4 Western Blotting

Buffers, nitrocellulose/PVDF membrane, gel, Whatman filter paper no. 3, Western blotting apparatus, power supply.

3.5 Antibody Staining

Blotting membrane, blocking solution, primary anti-phospho-specific antibody, secondary enzyme-labeled antibody, substrate, TBS-T.

4 Methods

4.1 Collection of Tissue

Seeds are germinated and grown in controlled conditions. For analysis of thylakoid phosphoproteins, different seedlings are rapidly frozen in liquid nitrogen, transferred in liquid nitrogen and stored at −80 °C until isolation of thylakoid membranes [34] (liquid nitrogen freezing is important so as to immediately stop any reaction in the samples).

4.2 Isolation of Thylakoid Membranes

4.2.1 Procedure

Thylakoid membranes are isolated according to modified protocol of Rintamaki et al., 1996 [34]. All procedures are performed under green safe light.

Take out frozen leaves and add ice-cold isolation buffer containing 50 mM HEPES-NaOH, pH 7.5, 300 mM sucrose, 5 mM $MgCl_2$, 1 mM-EDTA, 10 mM NaF (phosphatase inhibitor), and 1% (w/v) bovine serum albumin. Rapidly homogenize (mortar and pestle used needs to be cold) and filter through Mira cloth. Centrifuge at 1500xg for 4 min at 4 °C. Wash the pellet with ice-cold 10 mM HEPES-NaOH, pH 7.5, 5 mM sucrose, 5 mM $MgCl_2$, and 10 mM NaF and pellet again at 3000 × *g* for 5 min at 4 °C. Suspend thylakoid pellet in small amount of storage buffer of 10 mM HEPES-NaOH, pH 7.5, 100 mM sucrose, 5 mM NaCl, 10 mM NaF, and 10 mM $MgCl_2$ [34, 35] before freezing in liquid nitrogen. Protect all the preparations from light and keep preparations ice cold during the isolation. Aliquot isolated thylakoid membranes and store at −80 °C.

4.2.2 Precautions

The buffers used in the protocol must be supplemented with protease and phosphatase inhibitors as these would be helpful in preventing proteolytic cleavage and dephosphorylation reactions. Performing all steps on ice would slow all metabolic reactions leading to better outcomes.

1. All procedures should be protected from light.
2. The samples should be quickly transferred to liquid nitrogen.
3. Isolation should be done with ice-cold buffers at 4 °C.
4. Isolation buffer should have phosphatase inhibitor.
5. Any sample can be utilized depending upon requirement as per the standard protocols.
6. Protein should be aliquoted; one aliquot should be used only once.

4.3 Protein Estimation

For protein estimation, take out one aliquot of each sample and allow it to thaw on ice. Follow Bradford method of protein estimation (1976) [36].

4.3.1 Bradford Reagent Preparation

Prepare Bradford reagent by dissolving 40 mg of Coomassie blue-G in 20 mL of absolute alcohol with stirring for 20 min. Add 40 mL of ortho phosphoric acid and 340 mL of MilliQ water to the above solution. Filter the solution twice using Whatman filter paper no. 1 and store in dark bottle at 4 °C. All steps should be protected from light and steps should be performed with appropriate coverings wherever required.

4.3.2 Principle

Bradford method is a fast and sensitive method of protein estimation. This method is simple and suffers less interference by components of buffers. When a protein molecule binds with Coomassie dye in acidic conditions, it results in the change of its color from brown to blue. This relies on the reaction of basic amino acids as histidine, arginine, and lysine which are involved in formation of dye–protein complex. The higher the protein concentration, the higher is the intensity of blue color.

4.3.3 Procedure

Prepare standard curve by using different dilutions of BSA ranging from 1 μg to 20 μg. To 950 μL of Bradford reagent 50 μL of buffer or buffer-BSA or buffer-sample is added, vortex, and measure absorbance at 595 nm within 1 h.

4.3.4 Spectrophotometry

Spectrophotometric studies can be done on UV-visible double-beam spectrophotometer (readings of one experiment shown in the table as absorbance) (Fig. 3).

S. No.	Buffer (for dilution) in μL	Bradford reagent (in μL)	BSA standard (1 mg/mL) in μL
Blank	50	950	0.0
1.	49	950	1.0
2.	48	950	2.0
3.	46	950	4.0
4.	44	950	6.0
5.	42	950	8.0
6.	40	950	10.0
7.	38	950	12.0
8.	36	950	14.0
9.	34	950	16.0
10.	32	950	18.0
11.	30	950	20.0
Unknown	49	950	1.0

4.3.5 Expected Outcome

From the absorbance and concentration values, a standard curve can be easily prepared on a graph paper (Fig. 3). It is plotted with concentration on the *x*-axis and absorbance on the *y*-axis. Best precision line should be drawn for standards tested to find the unknown protein sample concentration.

BSA standard (1mg/mL) in microg	Absorbance (of experiment)
0	0
1	0.041
2	0.081
4	0.161
6	0.24
8	0.341
10	0.425
12	0.487
14	0.572
16	0.643
18	0.72
20	0.807

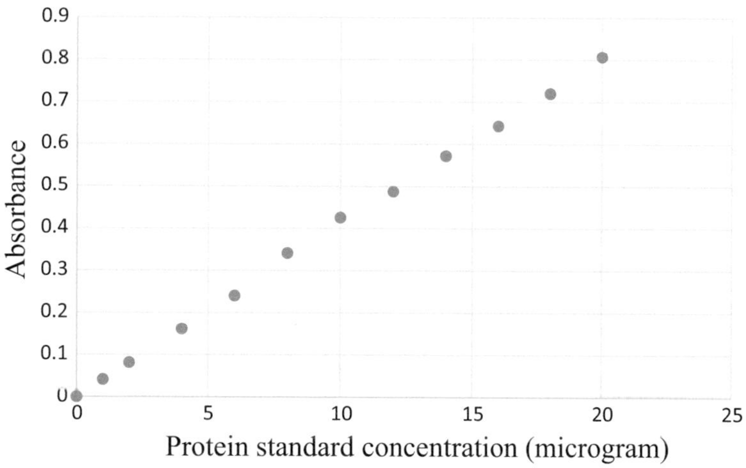

Fig. 3 Standard values-concentration of BSA standard with their respective absorbance. The concentrations of BSA (in microgram) are plotted against their absorbance (OD595) to prepare standard curve, which will help to determine concentration of unknown sample

4.3.6 Precautions

Pipetting should be accurate as minor errors may lead to uneven values.

1. Bradford is a sensitive technique and works well with lower protein concentration; therefore, absorbance beyond the standard range should be re-evaluated with diluted samples.
2. The reagent suffers interference in the presence of detergents as SDS in the sample (usually SDS is present in total protein extraction buffers).
3. Reading should be recorded in duplicate within the stipulated time.

4.4 Sodium Dodecyl Sulfate–Polyacrylamide Gel Electrophoresis (SDS-PAGE)

4.4.1 Principle

Sodium dodecyl sulfate–polyacrylamide gel electrophoresis is a powerful technique which is able to separate biomolecules on the basis of their size under an electric field. Polyacrylamide gel percentage affects pore size of gelling molecules and hence mobility of proteins. Therefore, molecules which are smaller in size migrate faster than big sized molecules. Bigger molecules face more friction and resistance from the gel matrix. In denaturing or SDS-PAGE, sodium dodecyl sulfate (or sodium lauryl sulfate) is used. It is a negatively charged detergent. Apart from SDS, sample buffer contains 2-mercaptoethanol, glycerol, Tris buffer, and bromophenol blue dye. SDS binds with amino acids at a constant molar ratio; due to strong negative–negative repulsion it prevents protein folding. 2-mercaptoethanol is a reducing agent which reduces disulfide linkages between cysteine molecules in the protein; glycerol

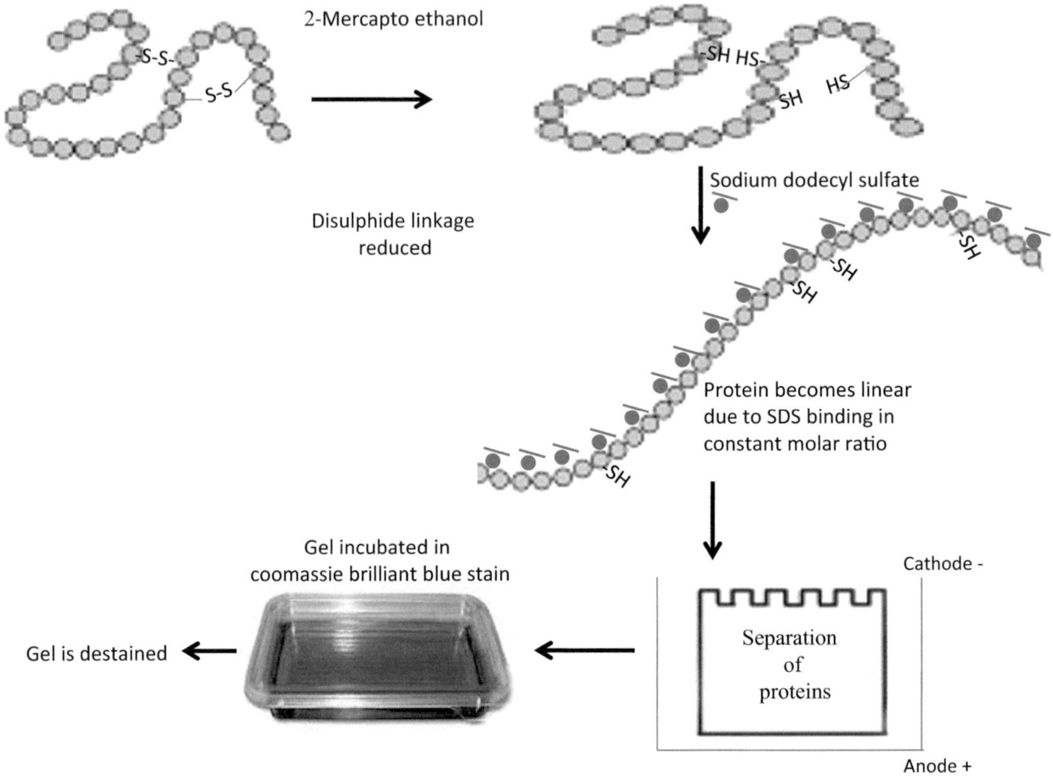

Fig. 4 Folded protein which is subsequently reduced by 2-mercapto ethanol and then denatured by SDS. Then it is electrophoresed after treatment with sample buffer. The running of the gel is tracked by bromophenol blue dye. After run is complete, the proteins can either be stained or processed for western blotting

provides density so that the sample stays at the bottom of the well; Tris is a buffering agent; and bromophenol blue present in the sample buffer is a tracking dye, which helps in visual tracking of the migration of molecules in the gel. Thus, in SDS-PAGE structure and charge of the proteins do not affect their migration but separation is affected by size of polypeptide linear chain (Fig. 4).

4.4.2 Procedure

Prepare monomer solution and other stock solutions of different constituents of gel as described by Laemmli [37]. Prepare monomer solution containing 30% acrylamide and 2.7% bisacrylamide. Store the solution in brown bottle at 4 °C. Prepare 4× separating gel buffer containing 1.5 M Tris pH 8.8. Prepare 4× stacking gel buffer containing 0.5 M Tris, pH 6.8. Prepare 10% SDS solution, and fresh solution of 10% ammonium peroxidisulfate. TEMED is used as such.

From these stock solutions, prepare 12.5% separating gel by casting mixture of 10 mL monomer solution, 6.0 mL 4× separating gel buffer, and 0.24 mL 10% SDS. Make the final volume to 24 mL with distilled water. Before casting add 200 μL of APS and 10 μL of TEMED to initiate polymerization and to stabilize cross-linking of monomer units in the gel. Pour the gel between two glass plates. Overlay 1 mL of water saturated n-butanol on to the separating gel to avoid air bubble. Allow it to polymerize for 2 h. Pour stacking gel onto the well-casted separating gel after removal of overlay solution. Add 1.7 mL of monomer solution, 4× stacking gel buffer, 1.0 mL 10% SDS, 150 μL of APS, and 5 μL of TEMED to a final volume of 10 mL with water for stacking gel. Pour the stacking gel mix after removal of overlaid water and insert the comb of desirable size. Allow it to polymerize for 1 h.

After polymerization remove the comb, wash the wells, and insert plates in the gel apparatus. The tank buffer/gel running buffer is added and the notched plate should be submerged in the buffer. The gel running buffer can be prepared by adding 0.025 M Tris pH 8.8, 0.192 M glycine, and 0.1% SDS. Solubilize thylakoid membrane proteins for in vivo phosphorylation in Laemmli sample buffer.

4.4.3 Protein Equalization

Load 20 μg protein in each well. All the samples are equalized for the protein concentration. Protein equalization is important as it affects ultimate quantitative evaluation. For all the samples, load a fixed amount of protein; for example, load 20 μg concentration for all the samples (if you have used 1 μL of sample for protein estimation, after concentration determination, calculate the amount required for 20 μg of sample). Determine the volume required from each sample for the appropriate protein concentration. Load the samples in the wells. In one lane, load medium range molecular weight markers (14.7–66.0 kDa) or other ranges of molecular weight standards depending upon the percentage of gel. In other wells, load thylakoid phosphoproteins isolated from different samples which have been solubilized in 6 M urea and sample buffer.

Run the gel either at constant voltage of 100 V or constant current of 20 mA.

4.4.4 Expected Outcome

The technique will separate the proteins on the basis of their size.

4.4.5 Precautions

1. Acrylamide solution should be stored in dark bottle.
2. Acrylamide is a neurotoxin and therefore should be handled with care.
3. Gel plates should be clean and properly sealed.
4. APS should be prepared fresh and should be added with TEMED prior to pouring.
5. Sufficient time should be given for casting of gel.
6. To avoid bubbles, overlay with water.
7. Spin protein sample before loading to avoid streaking due to particulate contaminants.

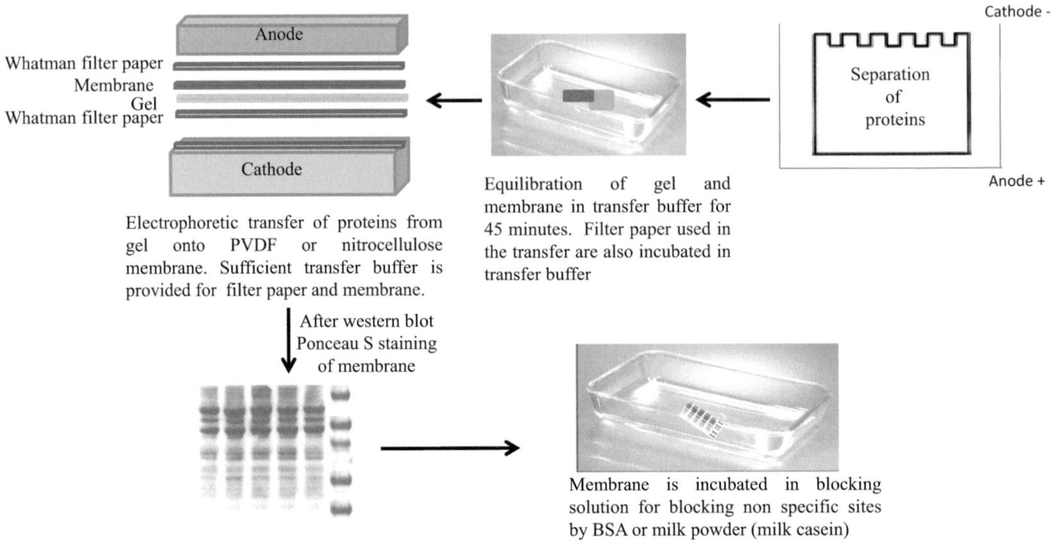

Fig. 5 Electrophoretic separation, equilibration of gel, membrane, and filter paper in transfer buffer for 45 min. After sufficient incubation, gel is placed toward cathode, then membrane is placed. At the top and bottom, few wet filter papers are placed and current is passed as described in materials and methods. After transfer, membrane is stained with Ponceau S for 2–5 min. Markers are marked by ball point pen and membrane is placed in blocking solution before antibody probing

4.5 Western Blotting

After the run, transfer polypeptides to nitrocellulose membrane by western blotting technique as shown in Fig. 5. Now stain the membrane briefly in Ponceau S stain (Ponceau S in acetic acid), you will see the markers and protein bands on the membrane. Mark the markers using ball pen as Ponceau S is rapidly washed off in subsequent steps. Now probe the membrane containing phosphoprotein with anti-phosphothreonine and anti-phosphoserine antibody by staining as described. After developing the membrane is carefully dried.

4.5.1 Principle

The proteins when separated by SDS-PAGE are electrophoretically transferred to nitrocellulose or PVDF membrane. The membrane is positively charged and suitable for protein transfer. It can sustain several treatments; therefore, it is easier to do remaining assays after transfer through western blotting. Nitrocellulose is weak but cheap as compared to PVDF membranes. PVDF requires prewetting and is strong with high absorption capacity.

4.5.2 Procedure

For western blotting, run SDS-PAGE as described in Fig. 5. Transfer the proteins to the nitrocellulose membrane (Fig. 5). After the run, equilibrate the gel in transfer buffer containing 0.1 M Tris, 0.192 M glycine, and 5% methanol v/v for 45 min. Cut exact sized nitrocellulose membrane and Whatman papers (measure the size of gel) and equilibrate them in the transfer buffer for 45 min. Add the buffer on transfer disk (anode). Place 4–6 pieces of 3 mm saturated

Whatman papers on the disk. Place gel on the papers and place the membrane carefully on the gel. Carefully remove air bubbles. Again place 4–6 wet Whatman papers. Apply constant current equal to twice the area of gel for 1.5 h during transfer. Handle the membrane only while wearing gloves.

After the transfer is over, mark the gel facing side of the membrane and stain it in ponceau S (0.1% ponceau S in 1% acetic acid).

4.5.3 Expected Outcomes

Staining shows the complete transfer. Mark the standard markers with ball point pen and wash the membrane with TBS-T/PBS-T to remove the stain.

4.5.4 Precautions

1. To determine the proper orientation, the gel and membrane should be given a small cut (may be at the bottom where markers were loaded) and always aligned to the cut. It is easy to check sequence of samples when markers are loaded, but in the absence of markers, it becomes difficult to acertain the orientation of the samples loaded onto the gel.
2. Gloves should be worn during all steps.
3. The size of membrane and filter sheets should be equal to that of the gel.
4. Ponceau S-stained membrane should be marked for markers with a ball point pen.

4.6 Detection of Phosphoproteins Using Antibody

4.6.1 Principle

Detection of proteins on membrane involves their immune staining by specific antibodies. The proteins separated by one-dimensional or two-dimensional electrophoresis after western blotting can be probed for the presence of specific proteins (after they are transferred to the membrane) (Fig. 6). The primary antibody is specific for protein and is usually unlabeled. Secondary antibody binds with primary antibody and is tagged with horseradish peroxidase (HRP) or alkaline phosphatase (AP) enzyme. Addition of appropriate substrate results in color development. The binding of enzyme labeled secondary antibody, which gives a signal, is proportional to the amount of primary antibody binding, which in turn depends upon the amount of the particular target protein in the sample. Thus, a strong signal suggests a high amount of target protein in the sample and a weak signal suggests a low amount of target protein in the sample. This technique is capable of detecting nanograms of protein and can be optimized for highly sensitive chemiluminescence detection.

4.6.2 Procedure

Keep the membrane in blocking solution containing 5% fat-free milk powder in TBST, at 37 °C for 1 h. TBST comprises 25 mM Tris pH 7.4, 136 mM NaCl, 0.2 mM KCl, and 0.05% Tween 20. After this, wash the membrane in TBST (TBST with Tween 20) thrice with constant shaking (each time for 5 min). Add primary

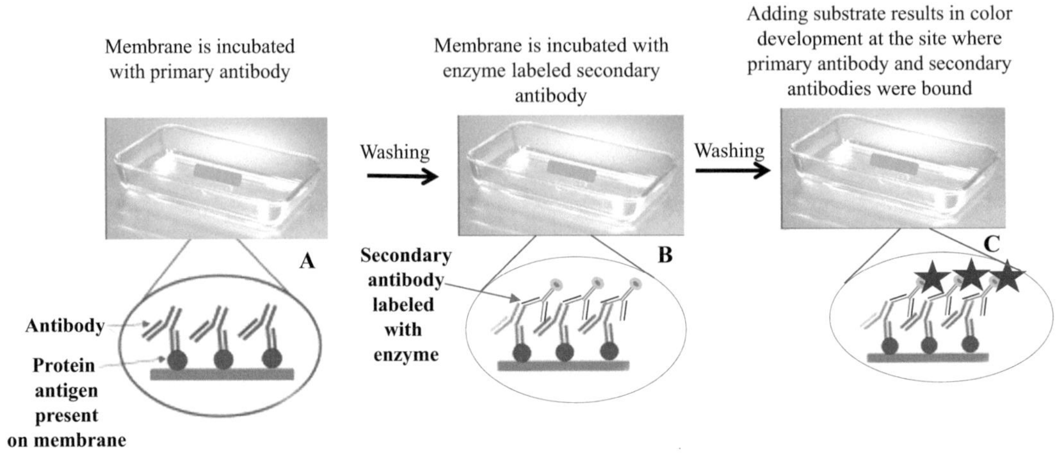

Fig. 6 (**a**) Probing of membrane with primary antibody. Primary antibody is specific for the protein antigen; in this case, it is against phosphothreonine (anti-phosphothreonine) or phosphoserine (anti-phosphoserine). (**b**) After incubation and washings, secondary antibody is added. Secondary antibody is labeled with enzyme alkaline phosphatase. (**c**) After incubation and washing, substrate is added which results in the formation of color bands at the target

antibody in appropriate dilution [Antibody dilution was done in 2.5% blocking solution with 0.05% sodium azide (as preservative)]. Incubate it for 1–2 h at 37 °C with constant shaking. Remove primary antibody (it can be stored at −20 °C for future usage). Wash the membrane three times with TBST for 5 min and then incubate with secondary antibody (diluted in TBS-T) for 1 h at 37 °C. Here, secondary antibody is tagged with alkaline phosphatase (ALP). Wash the membrane after incubation with TBST. Each time, washing is done to remove unbound antibody. After washing, substrate and buffer specific for the enzyme is added, which results in development of color. For alkaline phosphatase, 16 μL of 5-bromo-4-chloro-3-indolyl phosphate (BCIP) and 33 μL of nitro blue tetrazolium (NBT) are added to 5 mL of alkaline phosphatase buffer containing 100 mM Tris pH 9.5, 100 mM NaCl, 5 mM $MgCl_2$. Purple-blue bands would appear at the site of antibody binding. After staining the blot is washed in water and dried carefully (Fig. 7).

5 Notes

1. In case there is no development of color in protein assay, check the reagents and components of sample buffer and also explore any interfering agent present. In case your sample buffer has components that interfere with the protein estimation method, try to explore any other method of protein estimation. Or you can remove contaminants by protein precipitation by ammonium sulfate or trichloroacetic acid or acetone and resolubilize

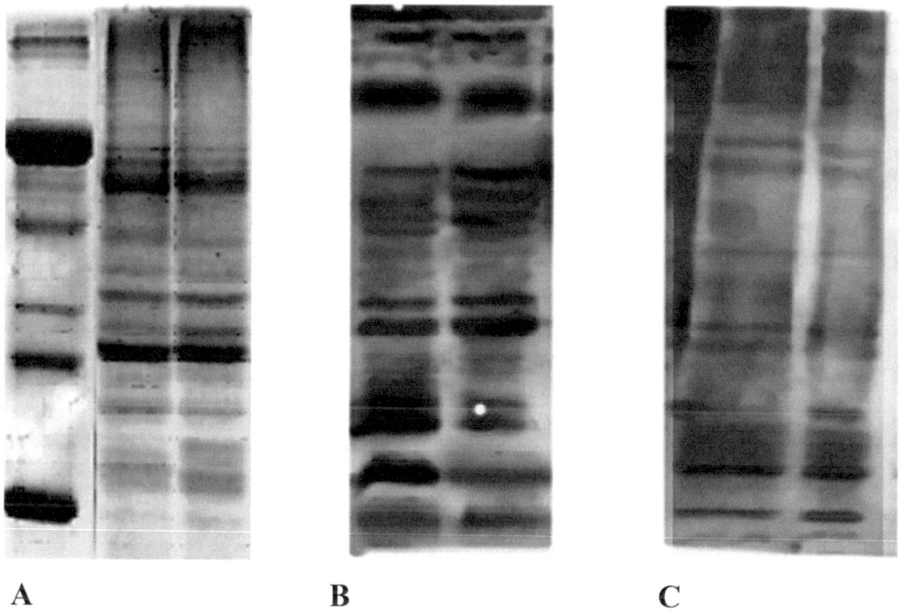

Fig. 7 (**a**) Thylakoid membrane proteins separated on SDS-PAGE. (**b**) Phosphothreonine specific phosphorylations in thylakoid membrane proteins. (**c**) Phosphoserine specific phosphorylations in thylakoid membrane proteins

in any compatible buffer, or you can try desalting and dialysis of the sample [38, 39].

2. Normally, the presence of detergents interferes with Coomassie reagents in Bradford method, while chelators, reducing agents, acids, and bases interfere with BCA and Lowry's method. Compatible buffers and accurate pipetting are key to successful protein estimation. However, color development depends upon many factors as the presence and abundance of particular amino acids, pH, buffer components, and so on [38, 39].
3. While separating your proteins on SDS-PAGE, you should make sure which molecular weight you intend to resolve; low, medium, or high molecular size would require high, intermediate, or low percentage acrylamide gels, respectively [38].
4. In SDS-PAGE depending upon your method of staining (Coomassie brilliant blue or silver staining or enzyme-based detection of antibody or radioisotopic labeling), the concentration should be chosen as all these detection methods offer different sensitivities. If your method is sensitive and protein concentration is high, it will result in high background and poor resolution [40].
5. During casting of gel use recommended quantities of ammonium persulfate and TEMED, using higher quantities would result in hard gel with poor resolution [39].

6. Streaking can be avoided by briefly centrifuging the protein sample to avoid any solid impurities. In case of smear formation, check your voltage for gel run. If bands are distorted, that indicates presence of salt; use any desalting method [38, 41].
7. Sometimes when buffers have contamination or acrylamide solution has problems, prepare fresh reagents, make proper solution, and if required degas before pouring of gel [42].
8. Heating effects should be avoided during runs. In winters, polymerization time is prolonged, try to cast gels at places with optimum temperature [42].
9. Carefully fill your upper and lower buffers in tanks. Avoid spillage of samples while loading, otherwise disturbed bands are observed on the gel. Prepare sample buffer carefully with all the components. Avoid using very old reagents [42].
10. In western blotting avoid using bare hands; always wear gloves while handling the membrane and gel. Always cut membrane and filter papers of appropriate size. Mark bands after staining with Ponceau S stain. Allow the incubation of membrane with each solution for the estimated time period. For example, diluted antibody would require prolonged exposure; likewise fresh and concentrated stock would require less time [42].
11. After developing you can carefully dry your membrane and store it.
12. All the steps of PAGE and blotting should be done as recommended by the user manuals of respective equipment.
13. During immune probing, for blocking of membrane, BSA should be preferred as milk may lead to unwanted background [43].
14. Immunoprobing and developing should be done carefully.
15. For cross-checking specific binding of the antibody, another lot of samples may be treated with phosphatase; if detection is not observed after their electrophoresis, blotting, and immune probing, it indicates that the antibody is highly specific.

References

1. Watson JC (2000) Light and protein kinases. Adv Bot Res 32:149–184
2. Luan S (2003) Protein phosphatases in plants. Annu Rev Plant Biol 54:63–92
3. Pawson T (1995) Protein modules and signaling networks. Nature 373:573–580
4. Johnson LN, Reilly MO (1996) Control by phosphorylation. Curr Opin Struct Biol 6:762–769
5. Tang G-Q, Hardin SC, Dewey R, Huber SC (2003) A novel C-terminal proteolytic processing of cytosolic pyruvate kinase, its phosphorylation and degradation by the proteasome in developing soybean seeds. Plant J 34:77–93
6. Huber SC, Hardin SC (2004) Numerous posttranslational modifications provide opportunities for the intricate regulation of metabolic enzymes at multiple levels. Curr Opin Plant Biol 7:318–322
7. Hardtke CS, Gohda K, Osterlund MT, Oyama T, Okada K, Deng X-W (2000) HY5 stability and activity in Arabidopsis is regulated

by phosphorylation in its COP1 binding domain. EMBO J 19:4997–5006
8. Neff MM, Fankhauser C, Chory J (2000) Light: an indicator of time and place. Genes Dev 14:257–271
9. Chen M, Chory J, Fankhauser C (2004) Light signal transduction in higher plants. Annu Rev Genet 38:87–117
10. Sullivan JA, Deng X-W (2003) From seed to seed: the role of photoreceptors in Arabidopsis development. Dev Biol 260:289–297
11. Quail PH, Boylan MT, Parks BM, Short TW, Xu Y, Wagner D (1995) Phytochromes: photosensory perception and signal transduction. Science 268:675–680
12. Chory J, Chatterjee M, Cook RK, Elich T, Fankhauser C et al (1996) From seed germination to flowering, light controls plant development via pigment phytochrome. Proc Natl Acad Sci U S A 93:12066–12071
13. Smith H (2000) Phytochromes and light signal perception by plants-an emerging synthesis. Nature 407:585–591
14. Quail PH (2002) Phytochrome photosensory signaling networks. Nat Rev Mol Cell Biol 3:85–93
15. Fankhauser C (2001) The phytochromes, a family of red/far-red absorbing photoreceptors. J Biol Chem 276:11453–11456
16. Cashmore AR, Jarillo JA, Wuand YJ, Liu D (1999) Cryptochromes: blue light receptors for plants and animals. Science 284:760–765
17. Lin C, Shalitin D (2003) Cryptochrome structure and signal transduction. Annu Rev Plant Biol 54:469–496
18. Huala E, Oeller PW, Liscum E, Han IS, Larsen E, Briggs WR (1997) Arabidopsis NPH1: a protein kinase with a putative redox sensing domain. Science 278:2120–2123
19. Briggs WR, Christie JM (2002) Phototropins 1 and 2: versatile plant blue-light receptors. Trends Plant Sci 7:204–210
20. Harter K, Frohnmeyer H, Kircher S, Kunkel T, Muhlbauerand S, Schäfer E (1994) Light induces rapid changes of the phosphorylation pattern in cytosol of evacuolated parsley protoplasts. Proc Natl Acad Sci U S A 91:5038–5042
21. Austin JR, Staehelin LA (2011) Three-dimensional architecture of grana and stroma thylakoids of higher plants as determined by electron tomography. Plant Physiol 155:1601–1611
22. Daum B, Kuhlbrandt W (2011) Electron tomography of plant thylakoid membranes. J Exp Bot 62:2393–2402
23. Barber J (1998) Photosystem two. Biochim Biophys Acta 1365:269–277
24. Hankamer B, Nield J, Zheleva D, Boekema E, Jansson S, Barber J (1997) Isolation and biochemical characterization of monomeric and dimeric photosystem II complexes from spinach and their relevance to the organization of photosystem II in vivo. Eur J Biochem 243:422–429
25. Andersson B, Anderson JM (1980) Lateral heterogeneity in the distribution of chlorophyll-protein complexes of the thylakoid membranes of spinach-chloroplasts. Biochim Biophys Acta 593:427–440
26. Nield J, Barber J (2006) Refinement of the structural model for the photosystem II supercomplex of higher plants. Biochim Biophys Acta 1757:353–361
27. Amunts A, Nelson N (2008) Functional organization of a plant photosystem I: evolution of a highly efficient photochemical machine. Plant Physiol Biochem 46:228–237
28. Aro EM, Suorsa M, Rokka A, Allahverdiyeva Y, Paakkarinen V, Saleem A et al (2005) Dynamics of photosystem II: a proteomic approach to thylakoid protein complexes. J Exp Bot 56:347–356
29. Barber J, Andersson B (1992) Too much of a good thing—light can be bad for photosynthesis. Trends Biochem Sci 17:61–66
30. Bonardi V, Pesaresi P, Becker T, Schleiff E, Wagner R, Pfannschmidt T et al (2005) Photosystem II core phosphorylation and photosynthetic acclimation require two different protein kinases. Nature 437:1179–1182
31. Wunder T, Xu WT, Liu QP, Wanner G, Leister D, Pribil M (2013) The major thylakoid protein kinases STN7 and STN8 revisited: effects of altered STN8 levels and regulatory specificities of the STN kinases. Front Plant Sci 4:417. https://doi.org/10.3389/fpls.2013.00417
32. Samol I, Shapiguzov A, Ingelsson B, Fucile G, Crevecoeur M, Vener AV et al (2012) Identification of a photosystem II phosphatase involved in light acclimation in Arabidopsis. Plant Cell 24:2596–2609
33. Gupta V, Tripathy BC (2010) Effect of light quality on chlorophyll accumulation and protein expression in wheat (*Triticum aestivum* L.) seedlings. Int J Biotech Biochem 4:521–536
34. Rintamaki E, Kettunen R, Aro EM (1996) Differential D1 dephosphorylation in functional and photodamaged photosystem II centers. Dephosphorylation is a prerequisite for degradation of damaged D1. J Biol Chem 271:14870–14875

35. Rintamaki E, Salonen M, Suoranta UM, Carlberg I, Andersson B, Aro EM (1997) Phosphorylation of light-harvesting complex II and photosystem II core proteins shows different irradiance-dependent regulation in vivo. Application of phosphothreonine antibodies to analysis of thylakoid phosphoproteins. J Biol Chem 272:30476–30482
36. Bradford MM (1976) A rapid and sensitive method for the quantitation of microgram quantities of protein utilizing the principle of protein-dye binding. Anal Biochem 72:248–254
37. Laemmli UK (1970) Cleavage of structural proteins during the assembly of the head of bacteriophage T4. Nature 227:680–685
38. http://www.bio-rad.com/webroot/web/pdf/lsr/literature/Bulletin_6201.pdf
39. www.sigmaaldrich.com/content/dam/sigma-aldrich/docs/Sigma/Bulletin/b6916bul.pdf
40. www.chem.ufl.edu/wp-content/uploads/sites/17/2014/05/SDS-PAGE-Stain-Protocol.pdf
41. www.hycultbiotech.com/media/wysiwyg/Troubleshooting_SDS-PAGE_1.pdf
42. Mahmood T et al. (2012) Western Blot: Technique, Theory, and Trouble Shooting. North Am J Med Sci 4(9):429–434
43. Yang PC, Mahmood T, (2012) Western blot: Technique, theory, and trouble shooting. North Am J Med Sci 4(9):429

Chapter 9

Estimation of Plasma Membrane and Tonoplast ATPase Activity in Plant Tissues

Neerja Srivastava

Abstract

ATPases are enzymes that hydrolyze adenosine triphosphate and play a significant role in the variety of cellular functions. They produce energy through hydrolysis of ATP for mechanical functions likes protein trafficking and degradation, transport of solutes, and cellular transfers. These membrane transporters perform various essential physiological functions and are very promising drug targets. ATPase hydrolytic activity is estimated to test enzyme functionality and also provides valuable knowledge of possible inhibitory influences of molecules which obstructs hydrolytic activity. The protocol discussed here is a basic method of isolating plasma membrane and tonoplast enriched fractions along with assay of plasma membrane and tonoplast ATPase activity. Proteins hydrolyze ATP in a reaction which releases inorganic phosphate which is then measured through a colorimetric assay. This protocol is highly adjustable and can be used to estimate ATPase activity in kinetic or endpoint assays. This protocol provides a basic framework for characterization of ATPases, is rapid and can be adjusted according to requirements.

Keywords Plasma membrane, Tonoplast, ATPases

1 Introduction

In biochemistry, the estimation of small quantities of inorganic phosphate is highly important. Phosphate estimation is especially helpful for assessing the hydrolytic rate of phosphatases which possess capacity to eliminate phosphate from their substrate through hydrolysis. The hydrolysis rate is linked with enzyme activity which is a highly significant functional parameter. In phosphatases, cation transporting adenosine triphosphatases (ATPases), generate inorganic phosphate through cleaving of γ-phosphate of ATP. These membrane transporters have several basic physiological functions and are promising drug targets. ATPase hydrolytic activity is estimated to investigate enzyme function. It also gives significant knowledge about potential prohibitory influence of molecules which meddle with hydrolysis [1].

Neelima Gupta and Varsha Gupta (eds.), *Experimental Protocols in Biotechnology*, Springer Protocols Handbooks, https://doi.org/10.1007/978-1-0716-0607-0_9,

Based on the current studies on ion uptake in plant roots, the primary active transport process of nutrients is considered to be an electrogenic transport of protons. The proton transport by a proton pump generates a proton gradient across the cell membrane, and the proton motive force involved in the gradient, controls the transport of ions and other solutes through the symport and antiport systems. Proton pumping may be mediated by membrane-associated ATPase [2]. Two types of membrane-associated ATPases in the plant root have been well characterized. The plasma membrane-type and the tonoplast-type H^+-ATPases were distinguished by vanadate and nitrate inhibition, respectively [3].

Proton pumping has been detected using membrane vesicles (microsomes, tonoplast, and plasma membrane vesicles). The vesicle is a useful model system to study ion transport in plant root. The characterization of the proton pump and ATPase in microsomal vesicles is important for studies on the nutrient uptake in relation to plant nutrition. Phenomenon of proton extrusion [4] and the localization of proton pumps along the roots of intact sunflower plants and various types of root-induced acidification of the rhizosphere were of considerable ecological importance for the plant-soil relationships in general and for the mobilization of mineral nutrients from components with low solubility [5].

Plant plasma membrane H^+-ATPase is a P-type ATPase connecting ATP hydrolysis with H^+ extrusion and therefore produces an electrochemical gradient across the plasmalemma. Proton gradient is required for secondary transport, cell elongation as well as maintenance of membrane potential. In this chapter, the readers would learn to estimate ATP hydrolytic activity of the plasma membrane and tonoplast-ATPase [6] from the plant source.

A characteristic feature of H^+-ATPase associated with plasma membrane is the ability to combine the hydrolysis of ATP with proton pumping across the membrane. Therefore, this activity can be estimated through the rate of ATP hydrolysis (enzymatic, colorimetric method) or as a proton transport (estimation of H^+ movement linked with the variations in acridine orange absorbance). It is generally apparent that a critical factor which determines the credible measurements of both activities is the quality and quantity of membranes [7]. Measurement of the plasma membrane H^+-ATPase activity is critical for knowledge of its function as well as regulatory process. But sometimes it is not easy to estimate the hydrolytic activity of the plasma membrane H^+-ATPase, as plant cells possess various ATP hydrolytic enzymes. In this protocol KNO_3 is used for inhibiting V-type ATPases and ammonium molybdate for acid phosphatases [8, 9]. Orthovanadate is an inhibitor for P-type ATPase and therefore can be employed to estimate the activity of the plasma membrane H^+-ATPase through the assessment of the liberated vanadate-sensitive Pi in ATP hydrolysis.

The Pi liberated in the reaction results in a blue complex which can then be estimated through absorption [6].

Two electrogenic H^+ pumps, the vacuolar type H^+-ATPase (V-ATPase) and the vacuolar pyrophosphatase, are present at membranes of the secretory pathway of plants. V-ATPase is the principal H^+ pump at endomembranes of majority of plant cells in terms of protein quantity as well as activity. The V-ATPase is essential for plant growth in normal situations, as it energizes secondary transport, maintains solute homeostasis, and probably assists in vesicle fusion. The existence of the cell is dependent upon maintenance or adjustment of the V-ATPase activity under various stresses such as salinity, drought, cold, acid stress, anoxia, and surplus heavy metals in the soil. Long- and short-term gene expression and activity are important in adaptation of the V-ATPase [10].

2 Technique

Various metabolic processes that need energy do so through coupling that process to the enzymatic hydrolysis of ATP to ADP and liberate free orthophosphate. This cellular process is performed through enzymes known as ATPases. The liberation of some energy possessed by ATP can be explained through the following reaction.

$$\text{ATP} + H_2O \rightarrow \text{ADP} + \text{Pi} + \text{Energy}$$

Enzymatic ATP hydrolysis is estimated through many different ways like measuring ADP through coupling enzymes, measurement of ^{32}P discharge from [γ-32P] ATP hydrolysis, or through colorimetric reactions. A majority of colorimetric reactions measure the discharged free orthophosphate (Pi) and are dependent on the formation of a phosphomolybdate complex in an acid medium after which a reduction or complexation with basic dyes takes place which forms colored complexes. The estimation of ATPase activity is dependent on measuring free phosphate from ATP.

3 Materials

3.1 Plant Material

Six-day-old peanut (*Arachis hypogea*, L) seedlings grown in Hoagland nutrient solution.

3.2 Reagents

3.2.1 Hoagland Nutrient Solution

Prepare nutrient solution according to Moore (1981) [11] with the following composition:

3.2.2 Stock Solution

Concentration	Chemical	Volume (mL)
1 M	Ca $(NO_3)_2$	10
1 M	KNO_3	10
1 M	$MgSO_4$	4
1 M	KH_2PO_4	2
	Fe EDTA[a]	2
	Micronutrients[b]	2

[a]5 mg metallic iron chelated with 37 mg Na_2EDTA to make 42 mg/mL solution of Fe-EDTA

[b]Micronutrient stock solution contains per liter 2.86 g H_3BO_3, 1.81 g $MnCl_2.4H_2O$, 0.11 g $ZnCl_2$, 0.05 g $CuCl_2.2H_2O$, and 0.025 g Na_2MoO_4

Mix all these stock solutions and make final volume 2.273 L with double distilled water.

3.2.3 Isolation Buffer

(a) Buffer for isolating Plasma membrane ATPase:

Reagent A) 50 mM Tris–HCl (pH 7.2), 0.25 M sucrose, 3 mM EDTA, and 0.04% 2-mercaptoethanol.

Reagent B) 25 mM Tris–HCl (pH 7.2) with 20% sucrose and 1 mM 2-mercaptoethanol.

(b) Buffers for isolating Tonoplast ATPase:

Reagent C) 20 mM Hepes-Tris (pH 7.6) with 250 mM sucrose, 1 mM Na_2EDTA, 10 mM $MgCl_2$, and 5 mM 2-mercaptoethanol.

Reagent D) 10 mM Hepes-Tris buffer (pH 7.0) containing 125 mM sucrose and 5 mM 2-mercaptoethanol.

4 Methods

4.1 Plant Cultivation

1. Soak the surface-sterilized peanut seeds (*Arachis hypogea L.*) in distilled water for 6 h in darkness at 25 °C.
2. Transfer the seedlings to the Hoagland nutrient solution.
3. Put the seedlings in a plant growth chamber in dark at 30 ± 20 °C and 80% relative humidity for 6 days.
4. Harvest 6-day-old peanut seedlings for isolating membrane fractions.

4.2 Isolation of PM Enriched Membrane Fraction

Carry out all the procedures at 0 °C, unless stated otherwise.

1. Take 25 g, 6 days old seedlings and wash them twice with distilled water.

2. Homogenize the seedlings in 100 mL of isolation buffer (Reagent A) consisting of 50 mm Tris–HCl (pH 7.2) with 0.25 M sucrose, 3 mm EDTA, and 0.04% 2-mercaptoethanol in a chilled mortar and pestle.
3. Filter the homogenate through four layers of cheesecloth.
4. Centrifuge the filtrate at 12,000 × *g* for 20 min using a fixed angle rotor and discard the pellet.
5. Centrifuge the supernatant at 105,000 × *g* for 60 min in an ultracentrifuge.
6. Resuspend the pellet in 4 mL of 25 mM Tris–HCl (pH 7.2) with 20% sucrose and 1 mM 2-mercaptoethanol (Reagent B).
7. Layer carefully this microsomal fraction on 5 mL of 34% sucrose in same buffer (Reagent B).
8. Centrifuge it at 80,000 × *g* in ultracentrifuge for 90 min.
9. Remove the clear sucrose layers by aspiration.
10. Collect the pellet containing plasma membrane.
11. Suspend it in Reagent B (25 mM Tris–HCl (pH 7.2) with 20% sucrose and 1 mM 2-mercaptoethanol).

4.3 Isolation of Tonoplast Membrane Enriched Fraction

Carry out all steps at 4 °C, unless stated otherwise.

1. Take 25 g of 6-day old seedlings and homogenize in 100 mL of Reagent C.
2. Filter the homogenate through nylon sieve (pore diameter ca. 500 μm).
3. Centrifuge the filtrate at 13,000 × *g* for 20 min using a fixed-angle rotor.
4. Discard the pellet and centrifuge the supernatant at 80,000 × *g* for 30 min using a fixed angle rotor.
5. Resuspend the pellet in 3 mL of reagent D and layer it over a 3-step gradient of 15%, 35%, and 45% sucrose (w/v) from top to bottom in reagent D without 125 mM sucrose.
6. Centrifuge it at 80,000 × *g* in a swing out rotor for 2 h.
7. Very carefully harvest tonoplast membrane enriched fraction at 15/35% sucrose interface through Pasteur pipette.

4.4 ATPase Assay

4.4.1 Buffers

Assay buffer: 50 mM Tris-MES (pH 8.0), 3 mM $MgSO_4$, 3 mM ATP, 50 mM KCl, and 100 μM ammonium molybdate.

Standard solution: 10 mM KH_2PO_4, diluted 1:10 to obtain 1 mM concentration.

Stopping reagent: 10% SDS.

Coloring Reagent: 6 part of 3.6 mM ammonium molybdate in 0.5 M H_2SO_4 + 1 part 10% ascorbic acid.

4.4.2 Assay Procedure

1. Analyze the tonoplast ATPase activity through the observation of release of Pi in the presence and absence of 50 mM NO_3^- ions. Add inhibitors of other ATPases like sodium azide (0.5 mM) for mitochondria and vanadate (0.2 mM) for plasmalemma in the reaction mixture.
2. Add 100 μM ammonium molybdate to inhibit acid phosphatase activity.
3. Determine the plasma membrane ATPase activity by observing the release of Pi in the presence and absence of 0.2 mM sodium orthovanadate.
4. Add 0.5 mM sodium azide, 50 mM KNO_3, and 100 μM ammonium molybdate in this reaction mixture to inhibit other ATPases and acid phosphatases.
5. Start the assay by adding membrane fractions (0.1 mL with about 2.0–2.5 μg protein) to reaction medium (0.45 mL) as given below:

   Reaction mixture.

	Probe (μL)	Standard (μL)	Blank (μL)
Assay buffer	300	300	300
Water	50	25	50
Standard	0	25	0
Membrane extract	100	100	100

6. Mix all the reagents, vortex the mixture immediately, and incubate for 45 min at 37 °C.
7. Terminate the reaction by adding 0.5 mL of 10% SDS.
8. Run blank and standards (with Pi) simultaneously, add 10% SDS prior to the addition of membrane fractions.
9. Determine the amount of Pi released [12, 13]. For this, add 0.6 mL of coloring reagent in the probe, standards and blanks and incubate at 370 °C for 45 min.
10. Keep tubes in ice for 5 min.
11. Read their absorbance at 820 nm.

4.4.3 Calculation of ATPase Activity

ATPase activity is expressed as ATPase activity/h/mL

$$\text{ATPase activity} = \frac{\text{OD of probe}}{\text{OD of standard}} \times \text{mmol of Pi} \times \text{enzyme dilution} \times \text{time}$$

Unit Definition: One unit of ATPase is the amount of enzyme that will generate 1.0 μmol of phosphate per min.

4.5 Conclusion

This is a very sensitive, rapid, consistent, and experimentally easy method for determining ATPase activity. This procedure can be easily used for estimating ATPase enzymes. The estimation of ATPase activity is not only essential to test enzyme functionality but also valuable to expose possible inhibitory influences of those molecules which hinder the process of ATP hydrolysis. Thus, this process is useful in biochemical and biomedical analysis of ATPase enzymes.

5 Notes

5.1 Precautions

1. Avoid foaming or bubbles while making buffers.
2. Take different tips for samples, standards, and reagents to evade contamination.
3. All reagents and solutions should be maintained at the suitable temperature before the start of the assay, and reagents should be added very carefully before use.
4. Always prepare fresh buffers before doing the assay.
5. Switch on all instruments and set them at suitable temperature before doing the assay.
6. Carefully follow the instructions, with proper incubation time and temperature according to the protocol.
7. Do not aliquot small volumes (<5 μL) as this will affect variability in results.
8. Use the correct wavelength and filter settings for the assay.
9. Follow the protocol carefully to identify the interfering substances as these can affect your results.

5.2 Avoiding Phosphate Contamination

Because assay estimates Pi liberation from ATP, any free Pi already existing in the assay mixture before the reaction starts will produce a background signal. Enzyme sample itself is the major source of free Pi, and all crude extracts of tissues and cells possess free phosphate unless some measures are initially taken to remove it. Free Pi in the enzyme may be removed through dialyzing or desalting.

Note: Phosphate buffer should never be used for preparation of tissue or cell homogenates.

5.3 Using Blanks Correctly

In general, assay blanks (i.e., lacking enzyme) should be prepared and the average blank value should be subtracted from all other tubes before doing any calculations with the assay data.

References

1. Bartolommei G, Moncelli MR, Tadini-Buoninsegni F (2013) A method to measure hydrolytic activity of adenosine triphosphatases (ATPases). PLoS One 8(3):e58615. https://doi.org/10.1371/journal.pone.0058615
2. Sze H (1984) H^+-translocating ATPases of the plasma membrane and tonoplast of plant cells. Physiol Plant 61:683–691
3. Sze H (1985) H^+-translocating ATPases: advances using membrane vesicles. Annu Rev Plant Physiol 36:175–208
4. Romheld V, Moll C, Masch H (1984) Localization and capacity of proton pumps in roots of intact sunflower plants. Plant Physiol 76:603–606
5. Wakiuchi N, Tanimoto H, Harada H, Oji Y, Shiga H (1988) Characterization of proton pumping and ATPase activities in microsomal fractions from barley roots. Soil Sci and Plant Nut 34(4):507–518. https://doi.org/10.1080/00380768.1988.10416467
6. Okumura M, Kinoshita T (2016) Measurement of ATP hydrolytic activity of plasma membrane H^+-ATPase from *Arabidopsis thaliana* leaves. BioProtocol 6(23):e2044
7. Janicka M, Wdowikowska A, Kłobus G (2018) Assay of plasma membrane H+-ATPase in plant tissues under abiotic stresses. Methods Mol Biol 1696:205–215. https://doi.org/10.1007/978-1-4939-7411-5_14
8. Shimazaki KI, Kondo N (1987) Plasma membrane H^+-ATPase in guard-cell protoplasts from *Vicia faba* L. Plant Cell Physiol 28 (5):893–900
9. Kinoshita T, Nishimura M, Shimazaki K (1995) Cytosolic concentration of Ca^{2+} regulates the plasma membrane H^+-ATPase in guard cells of fava bean. Plant Cell 7 (8):1333–1342
10. Dietz KJ, Tavakoli N, Kluge C, Mimura T, Sharma SS, Harris GC et al (2001) Significance of the V-type ATPase for the adaptation to stressful growth conditions and its regulation on the molecular and biochemical level. J Exp Bot 52(363):1969–1980
11. Moore TC (1981) Mineral nutrition in sunflower. In: Research experiences in plant physiology, a laboratory manual, 2nd edn. Springer-Verlag, New York
12. Leigh RA, Walker RR (1980a) A method for preventing sorbitol interference with the determination of inorganic phosphate. Anal Biochem 106(1):279–284
13. Leigh RA, Walker RR (1980b) ATPase and acid phosphatase activities associated with vacuoles isolated from storage roots of red beet (*Beta vulgaris* L.). Planta 150 (3):222–229

Chapter 10

Quantification of Conserved MicroRNA in Plants and Validation of New Targets

Lata Israni Shukla

Abstract

MicroRNAs (miRNAs) are approximately 20–24 nucleotide-long, posttranscriptionally regulatory RNAs. These arise from end nucleolytic processing of hairpin precursors. The plants perform transcriptional regulation by short interfering RNAs (siRNAs). The distinctive feature of miRNAs primarily lies in their biogenesis. The siRNA is processed from long double-stranded RNA while the miRNA is processed from single-stranded hairpins. The siRNA also has its role in transcriptional regulation instead of posttranscriptional regulation. The conserved miRNAs and their corresponding target genes are commonly found in all or most of the plant species. The highly conserved miRNA families are evolutionarily conserved across all major lineages of plants and play specific functions in a variety of physiological processes in plant species. These are rather important as they control the expression of important transcription factors by either cleaving or by translational suppression. The transcriptional factors control or regulate the expression of multiple downstream target genes by either activation or suppression. Thus, different regulation patterns of these conserved miRNAs imply complicated transcriptional and posttranscriptional regulation that determine the growth and development processes. The advantage in plant miRNA analysis lies in the positional nucleotidal preferences.

Keywords MicroRNAs, Stem-loop primers, DAMBE, psRNATarget

1 Introduction

The discovery of microRNAs (miRNAs) (21–24 nt length) and their interaction with the gene expression machinery in most organisms is one of the major scientific breakthroughs in recent years. The amount of miRNA with respect to total RNA ranges from 0.1% to 3% depending on the tissue type. Mediating the gene expression of cognate messenger RNAs (mRNAs) by nucleotide complementary pairing and assembly of RNA-induced silencing complex (RISC) are evidences for the regulatory effects of miRNAs. Evidences of plant miRNAs outside the cell from which they originate by cross-kingdom, intercellular exchange and recent reports of dietary plant microRNA in plasma and human targets in nucleotide complementary investigations are rather interesting.

Neelima Gupta and Varsha Gupta (eds.), *Experimental Protocols in Biotechnology*, Springer Protocols Handbooks,
https://doi.org/10.1007/978-1-0716-0607-0_10,

The detection of these microRNAs for quantification and novel miRNA identifications requires sensitive techniques. In this regard, the real-time quantification method is not only accurate and sensitive but also economical for the detection of plant miRNA.

microRNAs target the transcriptome in a nucleotide preferential manner; hence, it is rather important to identify and quantitate them; in silico identification and characterization is also done along with the wet lab techniques. Briefly for the wet lab technique the extraction of good-quality RNA is the first step, followed by the generation of single-strand cDNA with special stem-loop primers. Also, it requires the analysis of the prereported miRNAs for divergence and identification of conserved nucleotide positions. Since posttranscriptional regulation is conducted in a nucleotide-specific manner, statistical analysis and identification of new targets and nucleotidal preference are essential. Hence, programs like DAMBE and psRNATarget are routinely used.

2 Technique

2.1 Extraction of RNA

Nucleic acids are extracted from plant material in a manner that prevents degradation, as in plants the cells are covered with the cell wall. There are various key chemicals that play a vital role in the extraction of RNA from plant tissues. These chemicals are liquid nitrogen, guanidium thiocyanate, sodium acetate, 2-mercaptoethanol, phosphate buffer, and so on. The quantity of RNA extracted from roots, shoots, leaves, and stem are variable. The young seedlings, developing embryos, shoot tips, and root tips can yield high amounts of RNA. The procedure requires breaking the cell wall and cell membranes, by grinding using liquid nitrogen or solid carbon dioxide, keeping the tissue in frozen state and powdering it, followed by addition of two volumes of TRIzol in the mortar and pestle; it is allowed to thaw or the ground tissue could be transferred to a tube and extraction buffer is added to it. In order to purify RNA, insoluble particulates are removed through centrifugation, while soluble proteins and other material are separated by phase separation by phenol–chloroform–isoamyl alcohol treatment and centrifugation. One extraction with chloroform and isoamyl alcohol can further clear the aqueous solution that would separate from the phase separation method of a slight green tinge. RNA must then be precipitated from the aqueous phase and washed thoroughly to remove contaminating salts. The purified RNA is then resuspended in sterile distilled water. This method has been shown to provide intact RNA from plant tissue. To check the quality of the extracted RNA, a sample is run on an agarose gel, stained with ethidium bromide, and visualized under UV light.

3 Materials

3.1 Preparation of RNase-Free Water

RNase-free water is required for dissolving the purified RNA samples so as to avoid contamination with exogenous RNases.

RNase-free water
RNase-free water is the backbone of RNA extraction. It could be prepared by addition of 1 mL of DEPC (details given in the text) to 1 L milliQ water, uniformly dispersed and kept overnight (12 h). It is then autoclaved making sure no DEPC is visible as immiscible droplets after autoclaving
TRIzol preparation
For the preparation of TRIzol reagent, a highly toxic material guanidine thiocyanate (9.45 g) and ammonium thiocyanate (3.15 g) are used. These are dissolved in 50 mL of RNase-free water. To this solution, after uniform dispersion, distilled phenol (38 mL) is added which is followed by addition of 3.33 mL of 3 M sodium acetate solution and 5 mL of glycerol. The solution is mixed thoroughly at room temperature; the final volume is made to 100 mL. The solution can be stored in baked dark bottle at 4 °C for few months

MilliQ or a solution of 0.1% DEPC water (add diethylpyrocarbonate (DEPC) then leave it at 310 K overnight at room temperature and autoclave) can be used for extraction and further processing [1]. DEPC treatment is given to inactivate RNases. RNases break O–P phosphodiester bonds between RNA nucleotides by nucleophilic attack. In this type of reaction, a nucleophile (chemical group with a spare electron pair) attacks the partially positive charge of the O–P bond, resulting in phosphate and an OH group.

RNases are very diverse, but they usually have histidine, tyrosine, or serine as a nucleophile, which attacks and destroys bonds between nucleotides, degrading RNA molecules [2]. DEPC irreversibly reacts with the catalytic amino acid in the reactive center of the RNase molecule, inhibiting the enzyme [3]. It reacts most strongly with histidine residues but can also react with lysine, cysteine, and tyrosine residues, inactivating putative RNases in water. Excess DEPC is inactivated relatively simply by incubating at 310 K, which causes its hydrolysis, and this reaction is accelerated in the presence of nucleophiles such as amines, thiols, and alcohols [2–4]. Because Tris and HEPES are amines, instead of inactivating rare RNase molecules, DEPC safely fizzles out in most of these buffers

3.2 Preparation of the Plasticware and Glassware for RNA Extractions

It is essential to prepare RNase-free plasticware and glassware for RNA extractions. The plasticware are washed with dilute solutions of sodium hydroxide and then washed with sterile RNase-free water [1]. It can be made RNase free by covering open ends of glassware with aluminum foil and then baking at 432 K for 6 h.

It is essential that a specific area is marked in the laboratory which is dust free and not frequently visited by other colleagues and where

3.3 Preparation of Work Area for RNA Extraction

the person who is performing the extraction is not disturbed at the time of extraction. RNA can easily be degraded by saliva, sweat, RNases on fingertips, dust, etc.

3.4 Requirements for miRNA Expression Profiling

Primers

1. miRNA forward primer,
2. Universal reverse primer.
3. U6 forward primer.
4. U6 reverse primer.

Chemicals

1. RNase-free water.
2. SYBR Green I.
3. 6× DNA loading dye
4. 50 bp ladder
5. 20 bp ladder
6. TBE buffer.
7. Agarose.

Equipment

1. NanoDrop.
2. RT-PCR.
3. Horizontal gel apparatus.
4. Power supply.
5. Micropipette (1–200 μL).
6. Ice buckets.

Consumables

1. RNase- and DNase-free Microtips.
2. Gloves.
3. RT-PCR strip.
4. Autoclaved 0.5mL vial.
5. PCR tube stand.
6. Ice.

4 Methods

4.1 Laboratory Practice

Suitable gloves and mask for extraction of RNA in the allocated area in the laboratory is required. For a good RNA preparation, it is essential that the RNA extract is not contaminated with RNase

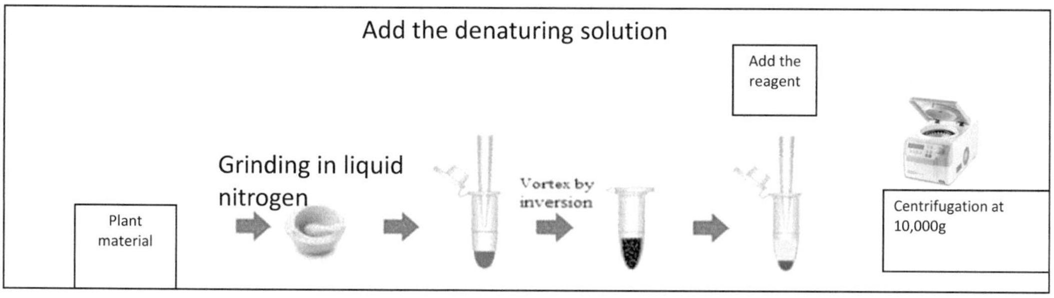

Fig. 1 Schematic diagram for the extraction of RNA from the plant tissue

from any source. The source of RNase includes the hands, saliva, and sweat. Care should be taken to obtain intact RNA. Repeated changes of gloves and utilization of RNase-and DNase-free tips, plasticware, and baked glassware are essential.

4.2 Grinding of the Plant Material

Fresh tissue is isolated from a particular part of a plant and placed in liquid nitrogen and stored at −80 °C till the time of extraction. Cut the tissue into small pieces (about to 100 mg), transfer to a mortar containing liquid nitrogen and pulverize the tissue using the pestle [1]. The liquid nitrogen has the special mode of action on these cells (Fig. 1).

Preparation of Denaturing solution

In 293 mL of DEPC-treated water, combine 250 g of guanidium thiocyanate, 17.6 mL of 0.75 M sodium citrate (pH 7.0), and 26.4 mL of 10% (w/v) sodium lauroyl sarcosinate. Guanidium thiocyanate is a strong denaturant; it disrupts the cell by solubilizing its components and denatures endogenous RNases. Mix the solution by magnetic bar with continuous heating at 337 K until all ingredients are dissolved. Store the solution at room temperature. Just before working, 0.36 mL of 14.4 M 2-mercaptoethanol per 50 mL of denaturing solution is added. For 100 mg of tissue, addition of 3 mL of the denaturing solution is required.

4.3 Homogenization and Centrifugation

The homogenate is transferred into a fresh polypropylene tube and 0.1 mL of 2 M sodium acetate (pH 4) is added, followed by 1 mL of phenol and 0.2 mL of chloroform–isoamyl alcohol per mL of the reaction mixture. Cap the tube and mix the contents thoroughly by inverting the tube (Fig. 2) [3].

Vortex the homogenate vigorously for 10 s and incubate the tube for 15 min on ice to permit the complete dissociation of the nucleoprotein complexes. For the phase separation centrifuge the

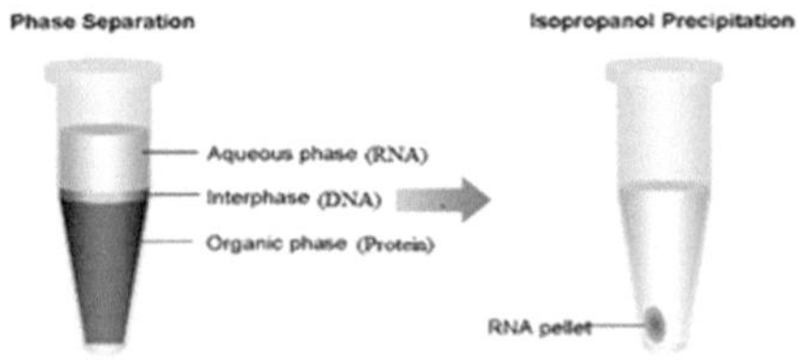

Phenol and Chloroform-isoamyl alcohol DNA carries a negative charge due to the phosphate groups which are neutralized in acid by protonation. On the other hand in RNA, is not neutralized in acid because, even though it also has a negative charge, it has exposed nitrogenous bases (it is single-stranded), which can form hydrogen bonds with water, keeping it in the aqueous phase. The phenol-chloroform combination reduces the partitioning of poly(A)+ mRNA into the organic phase and reduces the formation of insoluble RNA-protein complexes at the interphase. Due to their negatively charged phosphate backbone, nucleic acids are soluble in the upper aqueous phase instead of the lower organic phase (water is more polar than phenol) phenolic part contains the proteins and polymers including carbohydrates and lipids

Fig. 2 The principle of phase separation for separating RNA from tissue. The precipitation of RNA using isopropanol and high-speed spinning showing precipitation of RNA

tube at 10,000 × *g* for 20 min at 276 K [1]. Transfer the upper aqueous phase containing the extracted RNA to a fresh tube.

4.3.1 Precipitation

Add an equal volume of isopropanol to the extracted RNA, mix the solution well, and allow the RNA to precipitate for 1 h at 253 K. When compared with water, the isopropanol and ethanol have a much lower dielectric constant, making it much easier for Na^+ to interact with the phosphate group that shields its charge and makes the nucleic acid less hydrophilic, causing it to drop out of solution. Precipitation can also be done for extended period of time; however, this also influences the precipitation of salts and adversely affects the quality of the RNA. The RNA extraction process can be arrested and contents stored at 253 K. However, extra effort then needs to be made for washing of the pellets. The pellets obtained after centrifugation are washed with ice-cold 80% ethanol [4]. This should be avoided as RNA preparation is rich in rare RNA materials and the possibility of losing RNA in the process of careful washing cannot be ruled out. The RNA is then centrifuged at 10,000 × *g* for 30 min at 269 K. The removal of the isopropanol/wash with ethanol could be done by careful removal. The pellet is then air-dried for 30–45 min in sterile conditions. The pellet would appear translucent and could now be dissolved in RNAase-free water. The mixing or dissolution of the pellet is performed by taping eppendoff gently.

4.4 Quantitative and Qualitative Analysis of RNA

Transfer the solution to a microcentrifuge tube, vortex it well, and precipitate the RNA with 0.6 V to 1 volume of isopropanol for 1h at 252 K (at this stage the tubes can be stored at 4 °C for 12 h). Collect the precipitated RNA by centrifugation at maximum speed for 10 min at 277 K in a microcentrifuge [1]. Measure the absorbance at 260 and 280 nm.

An A260nm reading of 1.0 is equivalent to about 40 μg/mL of RNA. Absorbance at 260 nm is used to determine the RNA concentration in a solution. RNA has its absorption maximum at 260 nm and the ratio of the absorbance at 260 and 280 nm is used to assess the purity of RNA preparation. Pure RNA has an A260/A280 of 1.8–2.0.

4.5 Extraction of RNA by TRIzol Method

RNA can also be extracted using TRIzol. The procedure can be used on fresh or stored frozen material. After homogenization in liquid nitrogen, add TRIzol and transfer the homogenate to a fresh polypropylene tube and sequentially add 0.1 mL of 2 M sodium acetate, 1 mL of phenol, and 0.2 mL of chloroform–isoamyl alcohol (24:1) per mL of denaturing solution. Mix thoroughly by inversion and incubate on ice for 15 minutes. Centrifuge the tubes at 10,000 × *g* at 276 K for 20 min and transfer the supernatant into a fresh tube. Add equal volume of isopropanol to extract RNA, properly mix, and allow to precipitate at 253°K for 30 min. Then carefully decant the isopropanol and air-dry the pellet at room temperature in sterile conditions. Add 20–50 μL DEPC-treated water, estimate the concentration of RNA by measuring the absorbance at 260 nm, and store the RNA solution at 203 K for a maximum of 10 days. Currently many RNA isolation kits of different manufacturers such as Invitrogen, Ambion, and Merck are available on the market.

RNA isolation by TRIzol method (ref. life technologies)

- Weigh the sample.
- Grind in prechilled mortar and pestle with liquid nitrogen (approximately 100 mg to 1 g in 10–15 min) without allowing the sample to thaw.
- Add double the volume of TRIzol according to the sample weight.
- Keep aside for 5–10 min, then use the pestle to uniformly disperse the solution.
- The semisolid slurry should be transferred using baked RNAase-free spatula to centrifuge tubes of 0.5 to 0 2 mL (ice tray). Mix uniformly and add 0.2 volume of chloroform (200 μL/1 mL TRIzol). Mix by inverting the tube (20–25 times). Incubate on ice for 20 mins. **Note: Do not vortex**; weight the tube for equal balance.
- Centrifuge at 12,000 rpm for 20 min at 4 °C.
- Collect the aqueous phase in a separate tube (avoid pipette error).
- Add equal volumes of chloroform; weight the tube for equal balance.
- Centrifuge at 12,000 rpm for 20 min at 4 °C.
- Collect the aqueous phase in a separate tube (avoid pipette error).
- Add equal volume of isopropanol (100%) to the tube and store at −20 °C for at least 3 hrs.Centrifuge at 11,000 rpm for 15 min at 4 °C.
- Remove the isopropanol. **Note: Mark the pellet by marker.** Allow the pellet to dry. Dissolve in 0.1% DEPC water with gentle tapping.
- Absorbance at 260/280 ratio of 1.8 and quantification of RNA in sample by NanoDrop and analysis of quality by bioanalyzer

Further the stringent RNA quality and quantity check could be performed by Quibit® 2.0 Fluorometer (Thermofisher, USA). In a clear PCR tube, add 199 mL Qubit buffer, properly mix with 1 mL reagent dye and keep for 5 min in dark. Add 1 μL RNA sample to the tube and take the readings on Qubit Fluorometer. The 260/280 ratios of all the RNA samples used in this study should be greater than 2.0.

Agilent 2100 Bioanalyzer (Agilent Technologies, Palo Alto, CA, USA) is used for computing the RIN (RNA integrity number).

Take 550 μL Gel matrix.

Spin at 3700 RPM in a given column.

Aliquot 650 μL in each tube.

Add 1 μL RNA dye blue and vortex.

Briefly centrifuge at 13,000 RPM on RT for 10 min.

2 μL sample RNA is denatured at 70 °C for 2 min.

Add 9 μL Gel dye in Chip (RNA Nano Chip LOT#TK12BK20) matrix well G (Circular).

Close with syringe (from 1 mL to 0) and leave it for 1 min.

Release the syringe up to 1 mL.

High-quality RNA should have an RIN value greater than 7.

4.6 Generation of Single-Strand cDNA

The generation of single strand cDNA from RNA requires reverse transcriptase enzyme which requires primer and dNTPs in sufficient amounts under conducive conditions. For miRNA, special type of stem-loop primers are required which are able to identify the miRNA and selectively enable the generation of cDNA (Flowchart 1).

4.6.1 Primers

The sequences of miRNAs to be investigated are downloaded from miRBASE. For example, miR156l-3p, miR167d-5p, miR168a-5p, miR168a-3p, miR169a, miR166i-3p, miR408a-5p, miR164a,b,f, miR164c, miR164e, and miR398-3p are selected from the miR-Base. Some of the subfamilies of miRNA with same sequence at the 3′ end have a common stem-loop primer. For forward primer specific to miRNA sequence, the 5′ end is extended by 5–7 nucleotide to increase the melting temperature. Universal reverse primer is specifically designed for the stem-loop and used for amplification of all the miRNAs. The stem-loop primer and forward primer of the miRNAs (Fig. 3) are designed by miRNA Primer Design Tool, based on the UPL (Universal Probe Library) probes.

Primers for the transcriptomes (ARF8, NBS-LRR, argonaute-like protein and desiccation associated protein, Uclacyanin, and DNA-dependent RNA Pol II) can be designed using Primer-

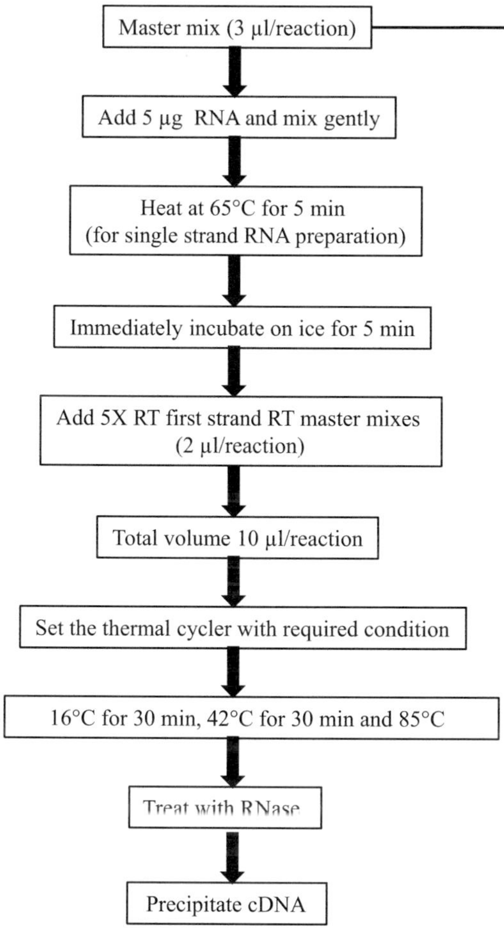

Reaction mixture	Concentration	1X
dNTPs	5 mM	1 µl
miRNA stem-loop primer	10µM	1 µl
U6 forward primer	10µM	1 µl

Flowchart 1 The schematic diagram for cDNA synthesis using 5× First-Strand RT master mixes (Takara)

BLAST of National Center for Biotechnology Information (NCBI) (Appendix 10)

Also, it is essential to perform an in silico validation of these primers' specificity by SnapGene software (from GSL Biotech; available at snapgene.com).

4.7 Generation of cDNA Using Stem-Loop Primers

The first step includes the hybridization of the stem-loop RT primer to a miRNA molecule which is then reverse-transcribed in a pulsed RT reaction. Later this reaction product is used for amplification and quantification using SYBR Green I assay.

The stem-loop primers are unique, and they enable the detection of small miRNA in a nucleotide-specific manner. The reverse transcription reaction is done in 10 µL containing the RNA, 50 nM stem-loop RT-primer along with a sufficient amount (0.25 mM) of each dNTP, also 50 units of reverse transcriptase, 1× reverse transcriptase buffer, 10 mM DTT, and 4 units of RNase inhibitor. The reaction is incubated for 30 mins at 289 K. This is followed by pulse

Primers		Sequence 5'-3'
U6	Forward primer	5'-ATTTGGACCATTTCTCGATTTGT-3'
	Reverse primer	5'-CGATAAAATTGGAACGATACAGA-3'
miR156l-3p	Stem loop primer	5'- GTCGTATCCAGTGCAGGGTCCGAGGTATTCGCACTGGATACGACGCTGAC-3'
	Forward primer	5'-GGCGGGCTCACTTCTCTTTCT-3'
miR164a-5p	Stem loop primer	5'-GTCGTATCCAGTGCAGGGTCCGAGGTATTCGCACTGGATACGACTGCACG -3'
	Forward primer	5'-GGCTGGAGAAGCAGGGCA-3'
miR164c-5p	Stem loop primer	5'-GTCGTATCCAGTGCAGGGTCCGAGGTATTCGCACTGGATACGACTGCACG -3'
	Forward primer	5'-CGCGTGGAGAAGCAGGGTA-3'
miR164e-5p	Stem loop primer	5'-GTCGTATCCAGTGCAGGGTCCGAGGTATTCGCACTGGATACGACCTCACG -3'
	Forward primer	5'-GGCTGGAGAAGCAGGGCA-3'
miR166i-3p	Stem loop primer	5'-GTCGTATCCAGTGCAGGGTCCGAGGTATTCGCACTGGATACGACGAGGAA -3'
	Forward primer	5'-GGCGTCGGATCAGGCTTCA-3'
miR167d-5p	Stem loop primer	5'-GTCGTATCCAGTGCAGGGTCCGAGGTATTCGCACTGGATACGACCAGATC-3'
	Forward primer	5'-TCCGTGAAGCTGCCAGCAT-3'
miR168a-5p	Stem loop primer	5'-GTCGTATCCAGTGCAGGGTCCGAGGTATTCGCACTGGATACGACGTCCCG-3'
	Forward primer	5'-TGGTTTCGCTTGGTGCAG-3'
miR168a-3p	Stem loop primer	5'-GTCGTATCCAGTGCAGGGTCCGAGGTATTCGCACTGGATACGACATTCAC -3'
	Forward primer	5'-TGTTTGATCCCGCCTTGC-3'
miR169a-5p	Stem loop primer	5'-GTCGTATCCAGTGCAGGGTCCGAGGTATTCGCACTGGATACGACTCGGCA -3'
	Forward primer	5'-CGGACCAGCCAAGGATGACT-3'
miR398a-3p	Stem loop primer	5'- GTCGTATCCAGTGCAGGGTCCGAGGTATTCGCACTGGATACGACAAGGGG-3'
	Forward primer	5'-GCGGCTTGTGTTCTCAGGTCA-3'
miR408a-3p	Stem loop primer	5'-GTCGTATCCAGTGCAGGGTCCGAGGTATTCGCACTGGATACGACGCCAGG-3'
	Forward primer	5'-CCGCTGCACTGCCTCTTC -3'

Fig. 3 The stem-loop, forward and reverse, primers of mature miRNAs miR156-3p, miR164a,c,e, miR166i-3p, miR167d-5p, miR168a-3p, 5p, miR169, miR398a-3p, miR408a-3p universal reverse primer 5'-CCAGTG-CAGGGTCCGAGGTA-3'

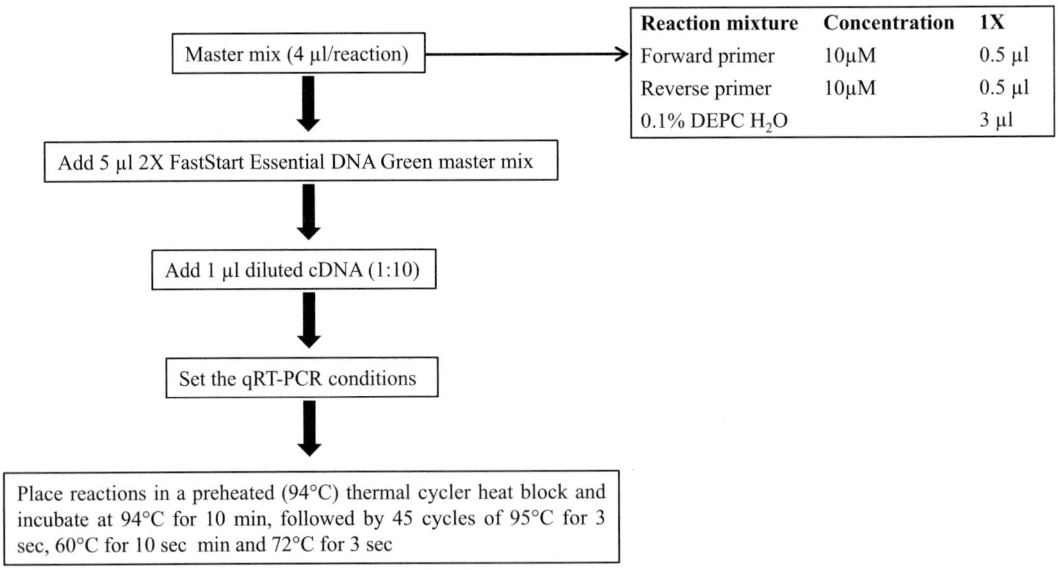

Flowchart 2 The schematic representation of qRT-PCR methodology

RT of 60 cycles at 303 K for 30s, 315 K for 30 s, and 323 K for 1 s. The pulsed RT reaction can then be terminated by incubation at 358 K for 5 min to inactivate the reverse transcriptase (Flowchart 2).

4.8 Designing of Primers

The 21 nt length of miRNA is too small to be successfully captured, so the primer for this can be selected from small hairpin where it binds and produces a product which is 150 to 200 bp long. This enables the increase in melting temperatures of the product followed by PCR using the reverse primers complementary to the stem-loop and a forward primer specific to the miRNA. Following the qPCR, the delta C_T value is determined and any housekeeping gene can be used as a reference.

4.9 RT-PCR of miRNA by Stem-Loop PCR

Plant microRNAs (miRNAs) play important roles in the posttranscriptional regulation of protein-coding genes, and they are essential for normal development and survival. The small size, the lack of a common features like a poly(A) tail, 3′ end-modifications, and the presence of a precursor—all these factors affect the detection and hinder the quantification of miRNAs. The stem-loop qRT-PCR method described here is designed to detect and quantify mature miRNAs in a fast, specific, accurate, and reliable manner. Firstly, a miRNA-specific stem-loop RT primer is hybridized to miRNA and then reverse-transcribed. Next, the RT product the single-strand cDNA is subjected to real time PCR analysis using a miRNA-specific forward primer and a universal reverse primer (Flowchart 3).

Step 1

Dilute prepared cDNA at 1:10 ratio.

Step 2

Evaluate the single-strand cDNA concentration by NanoDrop.

Step 3

Master mix preparation

1. miRNA (unknown sample) mater mix/sample,

Reaction mixture	Concentration	1× (single reaction) (μL)	Master mix 4× (μL)
miRNA forward primer	10 μM	0.5	2
	10 μM	0.5	2

(continued)

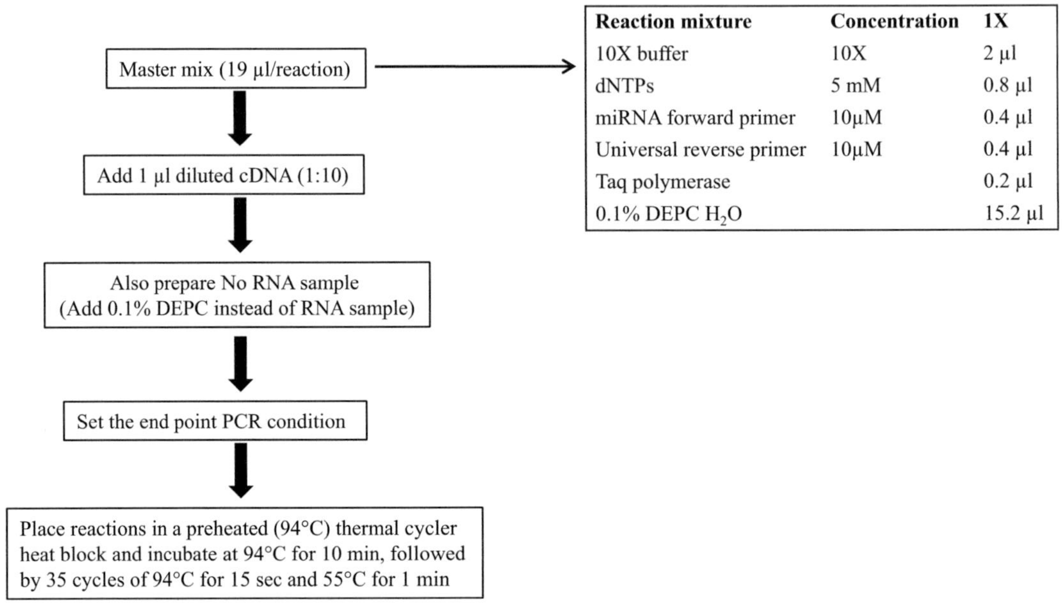

Flowchart 3 End-point PCR for detection of miRNA in the single strand stem loop cDNA mixture

Reaction mixture	Concentration	1× (single reaction) (μL)	Master mix 4× (μL)
Universal reverse primer			
cDNA	25 ng	0.5	2
DEPC H_2O	0.1%	3.5	14
Total volume		5	20

2. U6 primer master mix/sample.

Reaction mixture	Concentration	1× (single reaction) (μL)	Master mix 4× (μL)
U6 forward primer	10 μM	0.5	2
U6 reverse primer	10 μM	0.5	2
cDNA	25 ng	0.5	2
DEPC H_2O	0.1%	3.5	14
Total volume		5	20

3. No-template control master mix.

Reaction mixture	Concentration	1× (single reaction) (μL)	Master mix 4× (μL)
miRNA forward primer	10 μM	0.5	2
Universal reverse primer	10 μM	0.5	2
RNase-free water	–	4	16
Total volume		5	20

Step 4

Reaction preparation.

- Add 5 μL master mix per reaction.
- Add 5 μL SYBR Green I (in dark condition).
- Set the condition for RT-PCR.
- Condition for RT-PCR.

95 °C for 10 min—preincubation	
95 °C for 3 s	45 cycles
60 °C for 10 s	
72 °C for 3 s	

Step 5 Confirmation of product formation and the amount of primer dimers

RT-PCR product visualization (Flowchart 2)

- RT-PCR product gel electrophoresis.
- 4% agarose gel electrophoresis on 0.5% TBE buffer
- 5 μL sample with 1 μL 6× DNA loading dye
- 20 bp/50 bp ladder should be used.
- Picture taken after 1 h in gel doc (Fig. 4).

4.9.1 Retrieval of Mature miRNA Sequences

The miRBase database is a searchable online database of published miRNA sequences and annotation. The mature miRNAs sequence could be retrieved from miRBase ver.21 database (http://www.mirbase.org/) for reported plant species. The mature miRNAs originate from the miRNA stem-loop precursor (pre-miRNA) either 3′ or 5′ arm. Based on the origin from the precursor arms, miRNAs are categorized into 5′ and 3′ groups and investigated [5] (Fig. 5). The consensus sequences are generated using DAMBE.

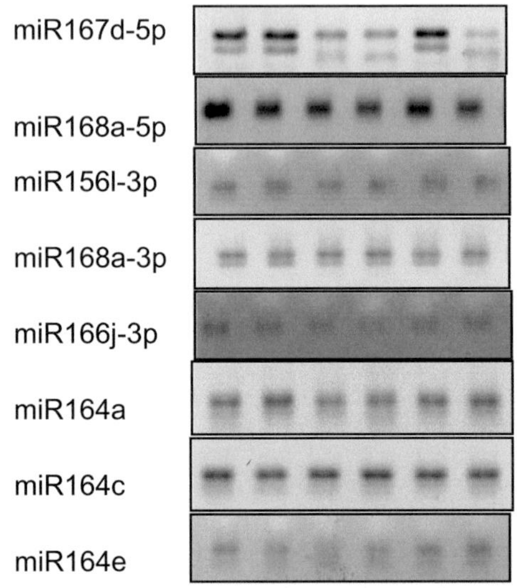

Fig. 4 End point PCR for the preliminary evaluation of miRNA in samples

4.9.2 Computational Identification of Targets for miRNA

The identification of new targets is feasible by psRNATarget, a plant small RNA target analysis server, which has two important analyses, reverse complementary corresponding to miRNA and target site accessibility evaluation. Targets from psRNATarget (http://plantgrn.noble.org/psRNATarget/) are predicted on expectation value 5.0 (Figs. 6 and 7).

4.9.3 Position Weight Matrix Analysis of miRNA

Position weight matrix (PWM) technique is used for characterizing a set of miRNA 5p and 3p belonging to different plant species (Fig. 6). The DAMBE v4.0.36 (Data Analysis in Molecular Biology and Evolution) software is used for the calculation of position weight matrix frequency of the categorized group of the unique targets. The FASTA file is provided to the DAMBE tool; a positive PWMij value is considered. A negative PWMij value means it is less likely to find nucleotide i (A/U/G/C) at position j than random expectation based on Pi values only. The PWM scores for each nucleotide and consensus sequences are then checked for 5p and 3p miRNA families; similarly PWM matrix is analyzed for the predicted targets by the miRNA-5p and -3p miRNA families (Isoforms) independently. The probability of encountering nucleotide of each group with site information is estimated. A comparison is made for the base sequences obtained for miRNA 5p families and the 5p families predicted targets.

Conservation positions and phylogenetic sequence similarities: Mature miRNA consensus sequences are used to generate WebLogo (a graphical representation of nucleotide at each position) using the online program WebLogo (v-2.8.2).

Analysis of consensus sequence of miR169-3p in 10 plant families comprising 92 sequences, the consensus sequence obtained for each plant family.

Families	Subfamilies	Plants	Sequences	1	2	3	4	5	6	7	8	9	10	11	12	13	14	15	16	17	18	19	20
All family	a-u	21	92	G	G	C	A	A	A	U	U	A	U	U	U	U	G	G	G	A	U	A	A
Brassicaceae	a-n	4	22	G	G	C	A	G	U	U	U	A	C	C	U	G	G	C	U	A	U	U	G
Cucurbitaceae	l	1	1	U	A	G	C	C	A	A	A	A	A	U	G	A	C	U	U	G	C	C	U
Euphorbiaceae	ad	1	1	U	C	A	C	A	G	G	C	U	C	U	U	A	U	U	U	U	U	C	A
Fabaceae	d,e,h,i,j,l,n,s	2	9	G	G	C	A	A	G	A	C	A	U	C	C	U	U	U	C	G	C	A	A
Linaceae	b	1	1	U	A	G	C	C	A	A	G	G	A	U	G	A	C	U	U	G	C	C	U
Amborellaceae	b	1	1	A	G	C	A	A	G	U	C	G	C	C	U	U	G	G	C	U	A	A	C
Poaceae	a-r	6	33	G	G	C	A	A	A	U	U	C	U	U	U	U	U	G	G	A	U	A	C
Rutaceae	c, d, e, l, n	1	5	C	A	G	G	C	A	G	U	C	U	C	C	U	U	G	G	C	U	A	A
Salicaceae	b,n,u	1	3	G	G	A	A	G	G	A	U	C	C	U	U	U	G	G	G	C	A	A	C
Solanaceae	a-g, m,o,s	3	16	C	A	G	C	C	A	A	G	G	A	U	G	A	C	U	U	G	C	C	G

Fig. 5 Positional preferences of the nucleotides (A,U,G,C) in the mature miRNA (5'–3'). Analyses across the subfamilies and the families were performed to generate consensus sequences

```
21 3'    C A G G G C U A G A C G U G G U U C G C U   5' 1      miR168a-5p
             : : : : : : : : : :   : : : : : :
638 5'   C C C C C G A U C U G C C C C A A G C U C   3' 658    Argonaute-like protein

21 3'    C A G G G C U A G A C G U G G U U C G C U   5' 1      miR168a-5p
           : : :   : : : . : : :   . : . : : : : :
1118 5'  C U C C G G A U U U G C U U C G A G C G A   3' 1138   Desiccation-associated protein
```

The *Os*-miR168a-5p targets Argonaute (AGO, validated; LOC_Os07g28850.1) and Desiccation associated protein (DAP, non validated; LOC_Os07g26480.1) obtained from the psRNATarget. Here, Colon (:) represent the perfect binding and dot (.) represent wobble pairing between the microRNA and target mRNA sequence. The miR168a-5p and AGO transcriptome (638 to 658 nucleotide) shows perfect binding is at 3rd-8th and 10th-19th positions. Whereas, binding between miR168a-5p and DAP transcriptome (1118 to 1138 nucleotide) shows perfect binding at 1st-5th, 7th, 10th-12th, 14th-16th and 18th-20th positions along with intermittent wobble pairing at 6th, 8th and 13th positions

Fig. 6 The miR168a-5p and target positional alignment red is target. Identified by psRNATarget and in black is the validated target argonaute-like protein

```
21 3'    A U C U A G U A C G A C C G U C G A A G U   5'  1     miR167d-5p
           : : : : : :     : : :   : : : . : : : :
728 5'   C A G A U C A A A C U G A C A G U U U C A   3'  748   NBS-LRR disease resistance protein

21 3'    A U C U A G U A C G A C C G U C G A A G U   5'  1     miR167d-5p
         : : : : : : :   : : : : : : : : : : :   :
2646 5'  U A G A U C A G G C U G G C A G C U U G U   3'  2666  Auxin response factor 8
```

The *Os*-miR167d-5p targets NBS-LRR disease resistance protein (non-validated; LOC_Os11g10570.1) and Auxin response factor 8 (validated; ARF, LOC_Os04g57610.3) obtained from the psRNATarget. Here, Colon (:) represent the perfect binding and dot (.) represent wobble pairing between the microRNA and target mRNA sequence. The miR167d-5p and NBS-LRR transcriptome (728 to 748 nucleotide) shows perfect binding is at 1st-4th, 6th-8th, 10th-12th and 15th-20th positions. Whereas, binding between miR167d-5p and auxin response factor 8 transcriptome (2646 to 2666 nucleotide) shows perfect binding at 1st, 3rd-13th and 15th-21st Positions.

Fig. 7 The miR167d-5p and target positional alignment, red is target. Identified by psRNATarget and in black is the validated target NBS-LRR disease resistant protein

4.9.4 Selection of Target

The selection of targets for the microRNAs is performed by using the sequence of the microRNAs for the miRBASE. The program for target identification includes psRNATarget. For example, the miR398 family targets are predicted using the expectation value 5.0

➤ The mature miRNA sequences were obtained from **miRBase** ***v 22*** **(2018)** through out the plant kingdom.

➤ The sequences were separated based on origin known as miRNA-3p and miRNA-5p in each plant families.

➤ The FASTA format of sequences were used for the position weight matrix analysis by using **DAMBE** software.

➤ Based on the nucleotide A, T, G, C positional preference identify the conservation of miRNA-3p and miRNA-5p.

➤ The miRNA-5p sequence of Poaceae family and *Oryza sativa were* chosen for target analysis.

➤ The "*Oryza sativa* (rice), transcript, TIGR genome cDNA OSA1 Release 5 (OSA1 R5), version 5" library with default settings was chosen in **psRNATarget** server (2017).

➤ Removed homologous and repeated targets sequences

➤ Based on seed region of miRNA-5p and target sequence complementary, the targets are divided in three groups (i) perfect complementary, (ii) near to perfect complimentary, (iii) imperfect complimentary

➤ **GC** content analysis in the target mRNA sequence aligned with miRNA sequence

➤ Identification of the predicted targets location in *Oryza sativa,* targets were subjected to **Chromosome Map Tool**

➤ The function categorization of targets were performed by **agriGO**

Flowchart 4 Representing various steps for mature miRNA analysis and identification of new targets

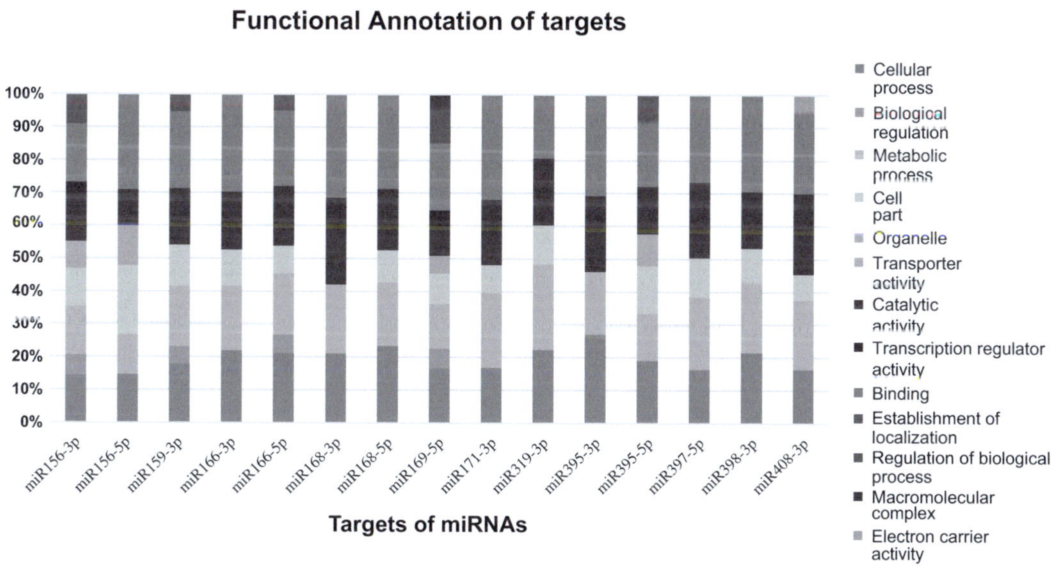

Fig. 8 The functional annotations of cleaved targets identified for the consensus sequence of the miRNAs using psRNATarget (2017 release), the functional annotations were performed using agriGo v2

by the psRNATarget (http://plantgrn.noble.org/psRNATarget/) (Flowchart 4). psRNATarget, a plant small RNA target analysis server, has two important analyses, reverse complementary corresponding to miRNA and target site accessibility evaluation [6].

4.9.5 Correlation with the microRNAs

The miRNA family analysis needs to be performed to identify the conserved positions that play a pivotal role in the binding to the targets. The utilization of the statistical technique using position weight matrix (PWM) analysis by using DAMBE for characterizing a set miR398 5p and 3p belonging to different plant species (Fig. 8). The DAMBE software (Data Analysis in Molecular Biology and Evolution) is used for calculating the position weight matrix frequency of the categorized group of unique targets [7–9]. The fasta file is provided to the DAMBE tool, version 4.0.36. PWM score analyzes the frequency distribution of the nucleotide; a positive score favors the nucleotide on the specific position, whereas a negative score suggests the lesser probability at the specific position. With the resulting PWM score for each nucleotide, a consensus sequence is obtained for miR398 5p and 3p miRNA families. Within plant families, the analysis is done for miR398 subfamilies also. The targets obtained in *Medicago truncatula* for miR398-5p and 3p are analyzed by PWM for miRNA-target nucleotide preferences. miRNA target binding regions (2–20 nt region) are analyzed. The probability of encountering nucleotide of each group with site information is estimated. A comparison is made for the base sequences obtained for miRNA 5p families and predicted targets. A similar approach is applied for the 3p family and predicted targets. The results are independently assessed for target nucleotide binding site polymorphism.

5 Notes

1. RNAs are susceptable to degradation due to RNAase activity. Therefore, its extraction and processing require tremendous precautions. Extraction requires own standardizations and care so as to maximize the yield and quality of RNA.
2. Casual experimentation without glassware and plasticware treatment may result in RNA degradation. Samples should always be aliquoted after RNA extraction.
3. DNA contamination in RNA extraction poses a serious problem; the phenol step needs to be carefully performed for minimizing DNA contaminants in the final preparation. DNase treatment and careful removal of the enzyme can be done in RNA extraction, before processing for cDNA synthesis.
4. If the ratio of 260/280 is low, it indicates protein contamination; if the ratio of 260/230 is low it indicates the presence of

inhibitors or guanidine salt. Ethanol wash and proper phase separation would be helpful for good-quality RNA.

5. Sometimes the amplification of miRNA may not be achieved because of low abundance. An increase in template would enable isolation of the miRNA.
6. Stem loop PCR for conserved miRNA in plants using the sequence provided herein is suitable.
7. Plant conserved miRNA show conserved positional preferences which inturn show target positional binding preferences. Using various tools the new targets could be identified.
8. Functional validation of the new targets could be performed by RT-PCR.
9. The plant miRNA biomarkers associated with plant growth and development could enable to make further advances in Plant biotechnology.

References

1. Chomczynski P, Sacchi N (2006) The single-step method of RNA isolation by acid guanidinium thiocyanate–phenol–chloroform extraction. Nat Protoc 2:581–585
2. John MC, Alan EP, MacDonald RJ, Rutter WJ (1997) Isolation of biologically active ribonucleic acid from sources enriched in ribonuclease. Am Chem Soc 18(24):5294–5297
3. Newbury HJ, Possingham JV (1977) Factors affecting the extraction of intact ribonucleic acid from plant tissues containing interfering phenolic compounds. Plant Physiol 60:543–547
4. Manuel GC, Francisco MC (1999) RNA isolation from plant tissues, a practical experience for biological undergraduates. Biochem Edu 27:110–113
5. Javed M, Solanki M, Sinha A, Shukla LI (2017) Position based nucleotide analysis of miR168 family in higher plants and its targets in mammalian transcripts. MicroRNA 6(2):136–142
6. Sinha A, Solanki M, Shukla LI (2019) Evidences for differential expression of miR167d-5p, target, positional nucleotide preference, and its role in somatic and different stages of regenerating calli of Oryza sativa. Plant Cell Tiss Organ Cult 136:537–548. https://doi.org/10.1007/s11240-018-01535-w
7. Solanki M, Sinha A, Shukla LI (2019) The miR408 expression in scutellum derived somatic embryos of Oryza sativa L. ssp. indica varieties: media and regenerating embryos. Plant Cell Tiss Organ Cult 136:537–548. https://doi.org/10.1007/s11240-018-01535-w
8. Solanki M, Sinha A, Shukla LI (2019) Optimization of in vitro culture media for improvement in yield of Navara ancient Indian medicinal rice. 3 Biotech 9(7):270. https://doi.org/10.1007/s13205-019-1797-2
9. Javed M, Sinha A, Shukla LI (2019) Evaluation of mature miR398 family, expression analysis and the post-transcriptional regulation evidence in gamma- irradiated and nitrogen-stressed Medicago sativa seedlings. Int J Radiat Biol 95:585–596

Chapter 11

Isolation of Natural Killer Cells from Mouse Bone Marrow and Spleen

Asmita Das

Abstract

Natural killer cells are large granular lymphocytes that are particularly important for killing virus infected cells and tumor transformed cells. Unlike T cells, NK cells are unique in their ability to kill target cells without prior sensitization. Hence, NK cells are very effective mediators for immune-modulation in combinatorial immunotherapy against tumors. The present chapter describes isolation of NK cells from spleen and bone marrow of mice and their purification. The chapter outlines a protocol for purification of NK cells isolated from spleen and bone marrow by depletion of contaminating cells in the preparation using negative selection process. Monoclonal antibody cocktail and magnetic separation is used to deplete contaminating cells. This process of negative selection has the advantage of the NK cells (cells being enriched for) not being coated with antibodies and hence not at risk of functional activation by antibody cross-linking and also helps in selecting all NK cells and not differential isolation of any skewed NK population. NK cells thus purified, have also been assessed for purity using flow cytometry by use of fluorescent tagged monoclonal antibodies for pan-NK markers. The functional potential of the NK cells isolated and purified have also been evaluated using cytotoxicity assay against prototypic tumor target cells.

Keywords NK cells, Negative selection, Flow cytometry, Chromium release assay

1 Introduction

NK cells are large granular lymphocyte that constitute 5–10% of peripheral blood lymphocytes. NK cells share common progenitors with T cells; however, they are classified as innate immune cells [1, 2]. NK cells are known to show changes in frequency of distribution and potency of activation in an organ specific manner [3]. They are effective in immunity against tumors and virus infected cells and can kill the target cells without prior sensitization. Many tumor cells downregulate their MHC class I expression in an attempt to escape cytolysis by cytotoxic T lymphocytes. However, such cells fall prey to NK cells as they have inhibitory receptors that engage MHC class I as their ligands and inhibit NK cell–mediated

Neelima Gupta and Varsha Gupta (eds.), *Experimental Protocols in Biotechnology*, Springer Protocols Handbooks, https://doi.org/10.1007/978-1-0716-0607-0_11, © Springer Science+Business Media, LLC, part of Springer Nature 2020

killing. Each NK cell expresses a multitude of activating and inhibitory receptors and the interplay of these opposing signals determines the outcome of the NK–target interaction [4]. Hence in the absence of the negative signal, activation signals win and result in cytolysis of the target cells [5, 6]. Thus, NK cells along with cytotoxic T lymphocytes and dendritic cells and macrophages play a very crucial role in immunity to tumors and virus infections [7]. Specialized NK cells are also found in the placenta and play an important role in successful pregnancy [8]. NK cells are also being implicated to play a crucial role in inflammation and autoimmune disorders [9] and even in tissue repair. Recent reports also exemplify a regulatory function of NK cells by direct cell to cell contact or via cytokines [10] and are particularly important in the context of viral, bacterial, protozoan diseases, cancer, autoimmune diseases and even in graft-versus-host rejections [11, 12].

2 Technique

NK cell holds the key toward devising effective therapeutic strategies in cancers, inflammation, bacterial and viral infections, and many other complicated clinical conditions. Isolation and enrichment of NK cells from primary and secondary lymphoid organs have been discussed along with the evaluation of NK cell activity, using cytokine secretion and cytotoxicity assays.

3 Materials

Mice and cell lines:C57Bl/6 mice: YAC 1 murine T cell lymphoma.

Chemicals and equipment: Dissection equipment including scissors, curved and straight forceps, teasers, mesh, syringe with plunger, dissection board. EasySep™ Mouse NK Cell Enrichment Kit through negative selection kit from Stem Cell Technologies along with magnet is used. Hemocytometer for cell counting and CO_2 incubator (Forma Scientific) and vertical laminar flow is used for cell culture using sterile plasticware from Costar Corning.

ELISA kits for IFN-γ and MIP1-α are procured from Invitrogen and chromium release assay kit from PerkinElmer. All antibodies (FITC anti-mouse NK1.1; APC anti-mouse CD3) for flow cytometry are procured from BD Pharmingen.

Specialized instruments: flow cytometer BD Biosciences; WIZARD® gamma counter; ELISA reader PerkinElmer.

Fig. 1 Experimental animal model (mouse)

4 Methods

Animals used are typically 6–8 weeks old, C57BL/6 mice weighing 25.0 ± 2.0 g (Fig. 1). All animals are maintained under controlled conditions of temperature (28.0 ± 5.0 °C), humidity (55%) and light 12:12 h of light and dark period. The animals are housed in polypropylene cages containing sterile paddy husk (procured locally) as bedding. Euthanization of animals is performed in accordance with the recommendations of the Institutional Animal Ethics Committee.

4.1 Use of Animal Model for Harvesting Mouse Spleen and Bone Marrow NK Cells

Spleen is the largest secondary lymphoid organ in the human body. NK cells share a common lineage with lymphoid cells. They reach the secondary lymphoid organ like spleen in mature naïve form and are ready to be activated by antigen in the secondary lymphoid organ. Hence NK cells isolated from spleen represent a mature NK population.

NK cells like other lymphoid cells develop in the bone marrow. The activation potential of mature NK cells is however adequate in bone marrow cells as well, as can be seen upon activation by cytokines like IL2. However, these cells have to be licensed by appropriate engagement of self MHC class I receptors [13].

4.2 Splenocyte Preparation

Aseptically remove the spleen from C57BL/6 mice and tease out in the recommended medium (PBS + 2% FBS + 1 mM EDTA). Pass the cell suspension through a 70 μm nylon mesh filter to remove unwanted debris and clumps and prepare a single-cell suspension. Resuspend the cell suspension in fresh recommended media. For Erythrocytes lysis, resuspend cells in 1mL of RBC lysis buffer, vortex for 2 min, and immediately add 1 mL of 2× PBS (Phosphate buffer salt concentration is double that of 1× PBS) to stop any further lysis by osmotic shock. Then wash the cells once in 1× PBS.

The NK cells thus isolated from splenocytes are further enriched using EasySep™ mouse NK cell enrichment kit through negative selection.

4.3 Bone Marrow Preparation

After euthanizing, dissect the mouse; pin the skin on a harvesting board. Harvest the femur and tibiae from the hind legs by cutting each end of the femurs to remove the hip and knee joints to expose the marrow using curved forceps. Carefully remove the muscle tissues adjoining the bone and keep the clean bones in a petri plate, making sure the bones are kept hydrated in PBS. Fill a 2 mL syringe with a 26-gauge needle with PBS. Using tweezers, the bone requires to be firmly held over a 15 mL centrifuge tube. Then insert the syringe needle into one end of the bone and depress the plunger to flush out the bone marrow. The syringe needle can be moved up and down inside the bone to flush out residual marrow. NK cells are then isolated from these bone marrow cells through EasySep™ Mouse NK Cell Enrichment Kit through negative selection kit.

4.4 Enrichment of NK Cells from Spleen and Bone Marrow Preparations

NK cells isolated from spleen and bone marrow are subjected to negative selection for enrichment of NK cells. The negative selection kit includes a cocktail of biotinylated antibodies containing a concoction of antibodies against cells other than NK cells typically present in preparations of mononuclear cells, like B cells, T cells, and monocytes. The kit therefore contains antibodies that target non-NK cells in the preparation and such cells are separated using streptavidin RapidSpheres (magnetic beads). When subjected to magnetic field, the non-NK cells therefore adhere to the wall of the tube due to the specific antibodies binding to the cells and are retained in the magnetic field and the cell suspension containing NK cells can be simply poured out into a fresh tube. The NK enrichment done by negative selection has the added advantage of not being nonspecifically activated by antibodies and can be directly used in subsequent experiments.

In brief, add 50 μL of NK cell enrichment cocktail to splenocytes cell suspension with concentration of 1×10^8 cells/mL. After mixing, incubate the cells at room temperature (RT) (18–25 °C) for 15 min. Further, add 200 μL of biotin selection cocktail, mix, and incubate for 15 min at RT. In the next step, add 200 μL of magnetic particles, mix, incubate for 10 min at RT, and bring the cell suspension to 2.5 mL volume by adding recommended media. Mix the cells gently by pipetting and place the tube into the magnet for 5 min. Pour the desired fraction in one continuous motion into the new tube and magnetically labeled unwanted cells will remain inside the original tube. The purity of NK cells is measured by staining cells with antibodies against NK cells marker (FITC anti-mouse NK1.1) and T-cell marker (APC anti-mouse CD3).

4.5 Evaluation of Cell Viability

The viability of NK cells and tumor cell lines is evaluated by trypan blue exclusion method [14]. In this method, 0.4% trypan blue dye is added to the cell suspension and the number of dead and live cells can be counted using hemocytometer. The live cells with intact membrane exclude the dye while dead cells are unable to do so, therefore, they appear blue in colour. The percentage of viable cells is counted by dividing the number of viable cells by the number of total cells counted and multiplying by 100.

$$\%\text{viable cells} = [\text{No.of live cells} - \text{No.of dead cells}/\text{Total no.of cells}] \times 100$$

4.6 Identification of Enriched NK Cells by Flow Cytometry

NK cells enriched by negative selection process are quantitatively evaluated by using pan-NK and pan-T cell markers that could stain all subpopulations of NK and T cells. Hence, the NK 1.1 marker is chosen for NK cells and CD3 marker is used for T cells. Greater than 90% enrichment is attained (representative flow cytometry results are shown).

Stain the NK cells isolated from spleen and bone marrow by pan-NK marker (NK1.1) to check for purity. Also, percentage of NKT cells are detected by pan-T marker (CD3). Resting or unstimulated NK cells have very limited activation potential and hence to efficiently study NK cell activity, the NK cells need to be stimulated using IL2, although resting NK cells also have activation potential. NK cells express intermediate affinity IL2 receptors and hence can be expanded using IL2. Adjust IL-2 stimulated and unstimulated NK cells to the concentration of 0.5×10^6 cells/mL and stain with anti-CD3 and anti-NK1.1. After mAb staining for 30 min at 4 °C, wash the cells and resuspend in ice-cold PBS buffer. All flow cytometry data are acquired using FACS LSR flow cytometer (BD Biosciences) (Fig. 2).

4.7 Testing NK Cell Activity

NK cells are large granular lymphocytes capable of killing virus infected cells and tumor cells without prior sensitization. NK cells have a multitude of activating and inhibitory receptors and the balance of opposing signals determines whether the target cell is spared or killed by the NK cells. Hence the killing of target cell is prevented by engagement of inhibitory receptors by cognate ligand. MHC class I serves as a cognate ligand for inhibitory receptors on NK cells. Hence, when some tumors downregulate the expression of their MHC class I, the engagement of inhibitory receptors by its cognate ligand is abrogated, leading to relieve of inhibition signal and resulting in activation of NK activity. YAC 1 (T cell lymphoma) tumor cells have very low MHC class I expression thus making them prototypic target cells for NK cells.

NK activity against YAC I target cells are tested by quantitative ELISA of the cytokines released, like IFN-γ and MIP1-α. NK-mediated cytotoxic activity can be checked by Chromium release assay [16, 17].

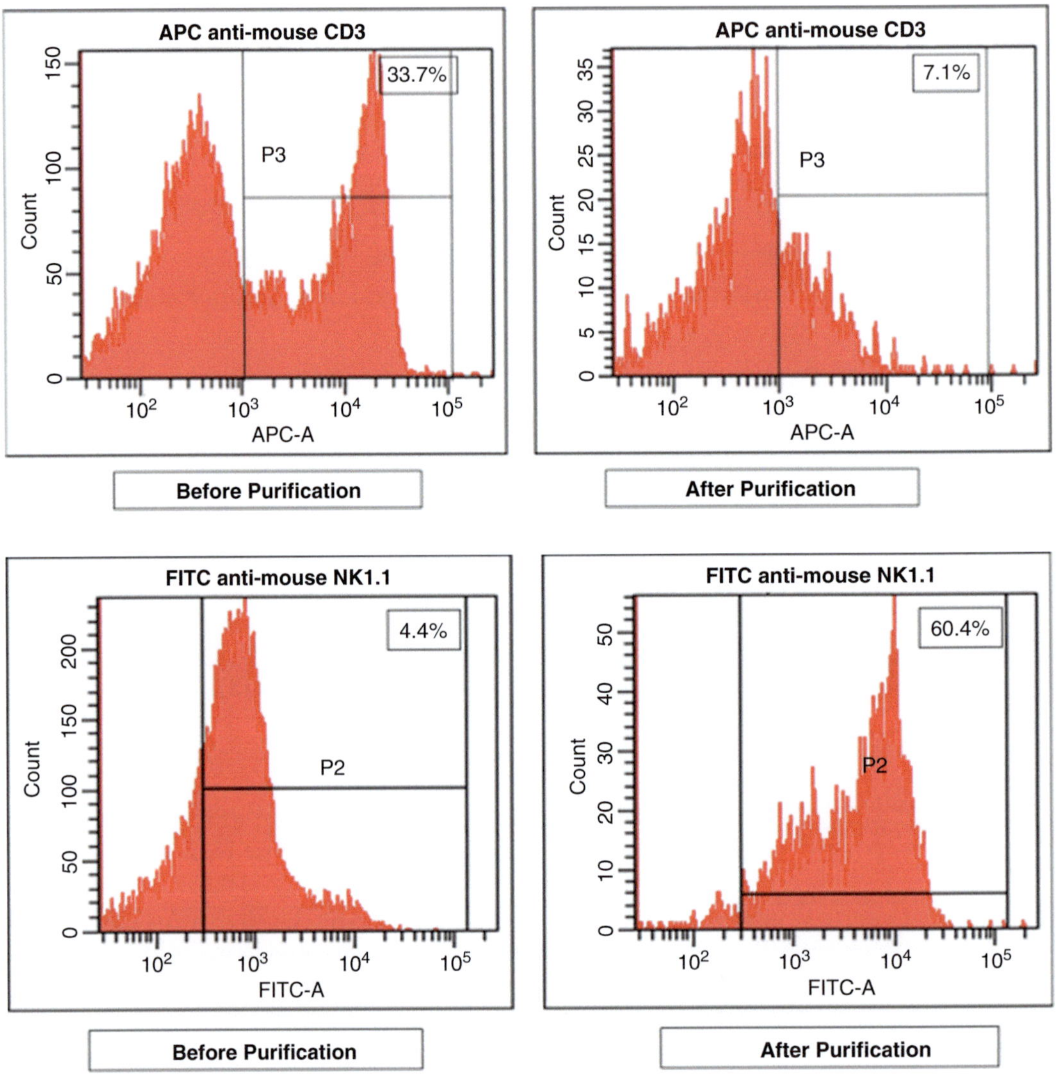

Fig. 2 Purification of NK cells. Spleenocytes subjected to purification (negative selection) were stained for $CD3^+$ cells (top panel) and $NK1.1^+$ cells (bottom panel). The cells were stained with APC anti-mouse CD3 and FITC anti-mouse NK1.1. Left panel shows population before purification and right panel show population after purification [15]

4.8 Preparation of Cells Culture Medium

Dissolve RPMI medium with 2 mM L-glutamine (supplied as dry powder) in 900 mL Milli-Q water and supplement with 25 mM $NaHCO_3$. Thereafter, add 10 mM HEPES, 1 mM Sodium pyruvate and antibiotics (viz., penicillin, streptomycin sulfate) and antimycotic (viz., nystatin) at a concentration of 50 Units/mL, 50 μg/mL and 2 μg/mL respectively. Adjust final pH to 7.4. Filter the medium using membrane filter assembly (0.22 μm) syringe. The sterility should be routinely checked by incubating the medium at

37 °C for 72 h. Store the medium thereafter at 4 °C. Prepare the final growth medium by addition of 10% FBS (heat inactivated) to the filter sterilized medium.

YAC-1 cells are routinely cultured and maintained as a monolayer culture in RPMI at 37 °C temperature under 5% CO_2 atmosphere and 95% humidity supplemented with 10% heat-inactivated FBS. Cells are passaged 2–3 times per week for maintaining them in healthy log phase.

4.9 Estimation of IFN-γ and MIP-1-α Cytokine Secretion Using ELISA

Quantitative estimation of cytokines is carried out according to the manufacturer's instructions using ELISA (R&D Systems). In brief, coat the plates by adding 100 μL/well of capture antibody to the 96-well plates. Seal the plates and incubate overnight at room temperature (RT). After completion of incubation period, aspirate the plates and wash three times with wash buffer by firmly tapping the plates upside down on absorbent paper. Block the plates by adding 300 μL/well blocking buffer and incubate at RT for 1 h. Thereafter, aspirate the plates and wash three times. Add 100 μL/well of standards and test samples to the coated plates. Seal the plates and incubate for 2 h at room temperature. Upon completion of incubation period, aspirate the samples and wash the plates three times with wash buffer. Add 100 μL of diluted detection antibody specific either to IFN-γ or MIP-1-α to each well. Thereafter, again wash the plates four times with wash buffer. Add 100 μL of diluted streptavidin–HRP solution (1:40). Wash the plates again four times with wash buffer. Then, add 100 μL of substrate solution (1:1 H_2O_2–tetramethylbenzidine (TMB) and incubate in dark for 20 min. Positive wells should turn bluish in color upon incubation with this step. Stop the chromogenic reaction by adding 50 μL of stop solution (2 N H_2SO_4) to each well. Positive wells should turn from blue to yellow. Read the final absorbance at 450 nm within 20 min of stopping reaction. Quantitative estimation of the expression of individual cytokine (IFN-γ, MIP-1-α) is carried out by extrapolation in terms of pg/mL of protein obtained from a standard curve prepared in parallel using standard supplied in the kit. A representative data clearly showing NK activation response upon coculture with YAC1 tumor cells has been shown in Fig. 3.

4.10 Chromium Release Assay for Detection of NK Cell–Mediated Cytotoxicity

Culture YAC 1 target cells in the presence of radioactive chromium, such that radioactive chromium is taken up by the tumor cells and sequestered inside the cells. When NK cells are cocultured with YAC 1 target cells, NK cells kill the YAC 1 target cells since NK inhibitory receptors are not engaged due to lack of MHC class I expression on YAC cells. Thus, upon killing of the target cells, chromium is released in the medium which can then be estimated after 4 h incubation at 37 °C using gamma counter or scintillation

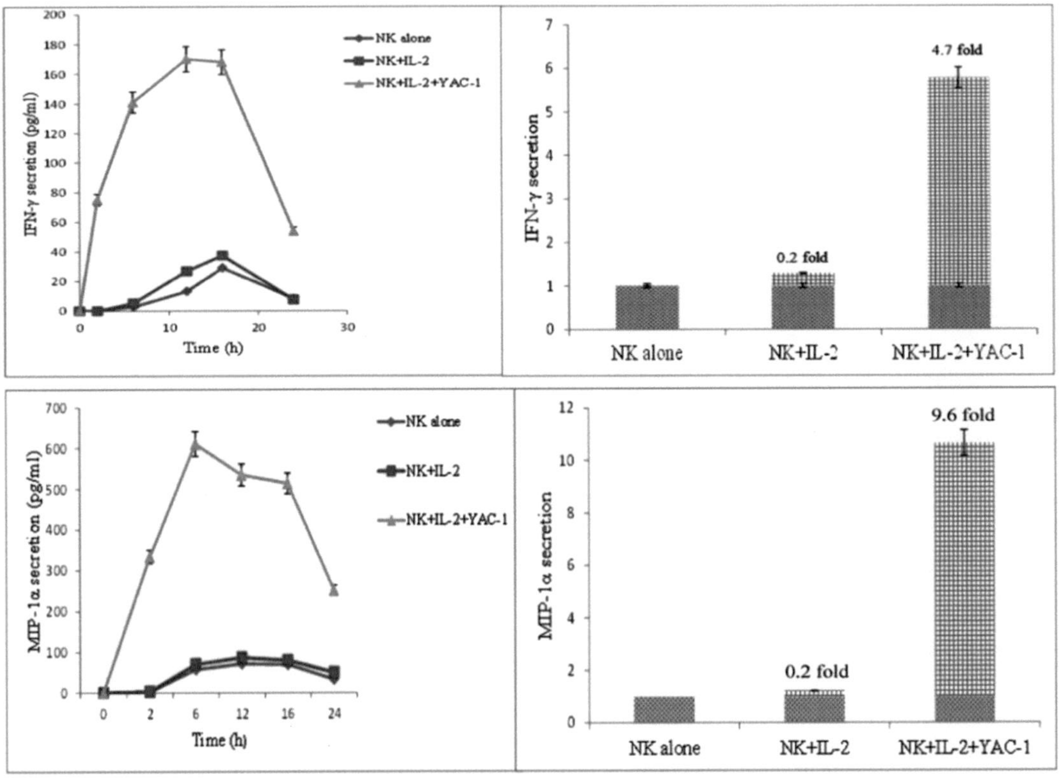

Fig. 3 Kinetics and profile of MIP1α and IFN-γ secretion on IL-2 stimulation. NK cells were incubated alone or with IL-2 or with YAC-1 cells as positive control at 37 °C. Supernatants were harvested at different time points and as indicated MIP1α and IFN-γ secretion were determined by using ELISA kit. Values represent mean ± SD of 3 independent experiments [15]

counter (as per PerkinElmer protocol) and is directly proportional to the cytotoxic activity of the NK cells. Percent cytotoxicity is calculated from data of triplicate wells (Fig. 4). Different Effector: Target ratio is used for data evaluation and statistical analysis [18].

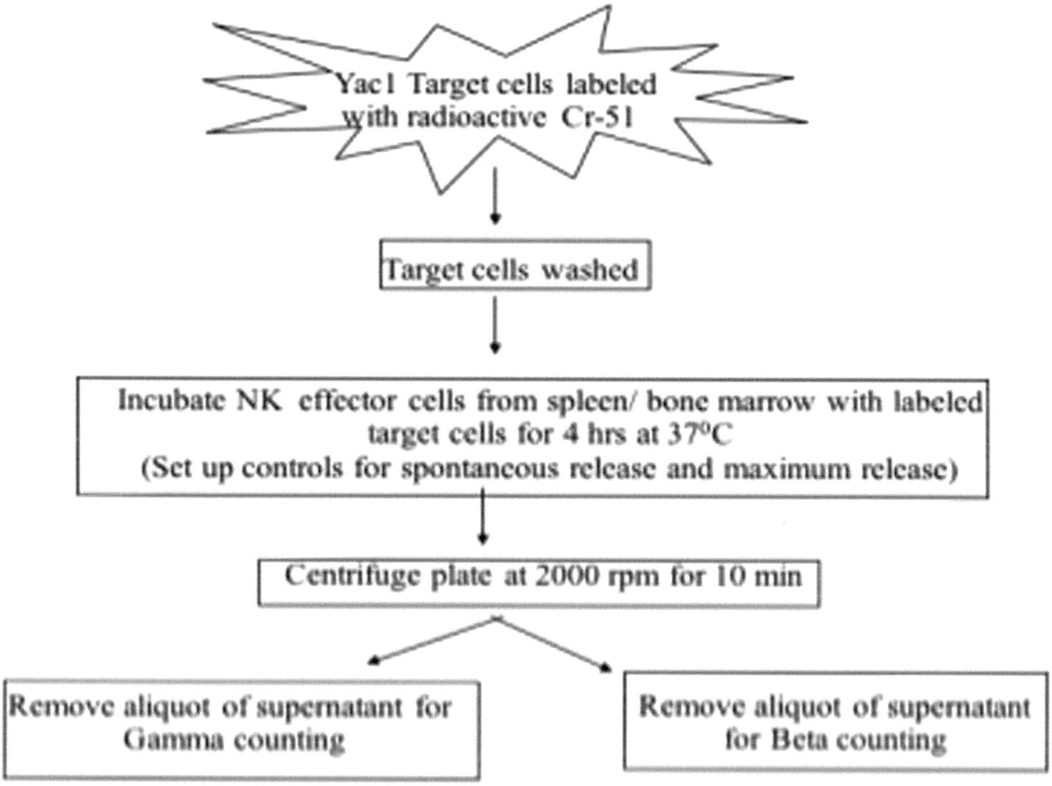

Fig. 4 Flow chart for chromium release assay

5 Notes

1. The present chapter focuses on the isolation of murine NK cells from spleen and bone marrow and the enrichment of NK cells from lymphocyte preparation by negative selection, evaluation of the cells thus enriched using the technique of flowcytometry. We have also shown the activation potential of these NK cells by cytokine production and cytotoxicity assays. NK cells isolated from primary lymphoid organs are mature but naïve, while NK cells isolated from spleen are mature activated NK cells that are already licensed and hence potentially active. The activating potential of NK have however been shown to be different in different organs and have even exhibited modulation particularly in tumor microenvironment.

   NK cells represent a population of large granular lymphocytes with the capability of killing tumor cells and virus infected cells without prior sensitization. NK cells, although have been classified as innate immune cells, have recently been shown to have many unique features like short-term memory response to viral infections [19]. NK cells have been discovered to play new roles in inflammation response, bacterial infections, autoimmunity [20] and retention of pregnancy, and many other roles.

References

1. Langers I, Renoux VM, Thiry M, Delvenne P, Jacobs N (2012) Natural killer cells: role in local tumor growth and metastasis. Biologics 6:73–82
2. Jiao Y, Huntington ND, Belz GT, Seillet C (2016) Type 1 innate lymphoid cell biology: lessons learnt from natural killer cells. Front Immunol 7:426
3. Sharma R, Das A (2014) Organ—specific phenotypic and functional features of NK cells in humans. Immunol Res 58(1):125–131
4. Vivier E, Raulet DH, Moretta A, Caligiuri MA, Zitvogel L, Lanier LL et al (2011) Innate or adaptive immunity? E example of natural killer cells. Science 331:44–49
5. Lanier LL (2008) Up on the tightrope: natural killer cell activation and inhibition. Nat Immunol 9:495–502
6. Lanier LL (2009) DAP10- and DAP12-associated receptors in innate immunity. Immunol Rev 227:150–160
7. Long EO (2007) Ready for prime time: NK cell priming by dendritic cells. Immunity 26:385–387
8. Moffett A, Hiby SE (2007) How does the maternal immune system contribute to the development of pre-eclampsia? Placenta 28: S51–S56
9. Poggi A, Zocchi MR (2014) NK cell autoreactivity and autoimmune diseases. Front Immunol 5:27
10. Laroni A, Gandhi R, Beynon V, Weiner HL (2011) IL-27 imparts immunoregulatory function to human NK cell subsets. PLoS One 6 (10):e26173
11. Souza-Fonseca-Guimaraes F, Adib-Conquy M, Cavaillon JM (2012) Natural killer (NK) cells in antibacterial innate immunity: angels or devils? Mol Med 18:270–285
12. Pallmer K, Oxenius A (2016) Recognition and regulation of T cells by NK cells. Front Immunol 7:251
13. Das A, Saxena RK (2004) Role of interaction between Ly49 receptors and cognate MHC I molecules in IL2 induced development of NK cells in murine bone marrow cell cultures. Immunol Lett 94(3):209–214
14. Strober W (2001) Trypan blue exclusion test of cell viability. Current Protoc Immunol 111:A3. B.1-3
15. Sharma R, Das A (2018) IL-2 mediates NK cell proliferation but not hyperactivity. Immunol Res. 66(1):151–157
16. Fu B, Tian Z, Wei H (2014) Subsets of human natural killer cells and their regulatory effects. Immunology 141(4):483–489
17. Paolini R, Bernardini G, Molfetta R, Santoni A (2015) Cytokine Growth Factor Rev 26 (2):113–120
18. Li Q (2018) Natural killer (NK) cell assays in immunotoxicity testing. Methods Mol Biol 1803:231–241
19. Lam VC, Lanier L (2017) NK cells in host responses to viral infections. Curr Opin Immunol 44:43–51
20. Abel AM, Yang C, Thakar MS, Malarkannan S (2018) Natural killer cells: development, Maturation and Clinical Utilization. Front Immunol 9:1869

Chapter 12

Gene Expression Analysis from Human Peripheral Blood Mononuclear Cells

Sipahee Lal Patel, Dinesh Kumar, Anamika Dwivedi, Payal Singh Raghuvanshi, Vishal Chand, Jaya Prakash, and Varsha Gupta

Abstract

Transcriptomic analysis or gene expression profiling is gaining importance in molecular biology research. Transcriptomics deals with studying the transcriptome or complete set of RNA transcripts produced in the specific cell at a particular time under some specific condition using high-throughput (DNA microarray) and low through put screening methods. In cells, DNA is the hereditary material which not only transfers genetic information from one generation to another but also codes for major proteins required for cell survival and communication. DNA is transcribed to RNA (mRNA is produced after processing of nascent or hnRNA) and mRNA then translates into proteins. The process of transcription and translation are referred as central dogma. Thus, the analysis of expression of genes can be done at either the level of transcript (analyzing RNA) or at the level of protein (analysis of proteins). In multicellular organisms, expression of genes and thus of proteins is under spatial (different in different cells and tissues) and temporal (different expression at different times) regulation. In transcriptomic/transcript analysis it is important to choose correct source and nature of the product, the resolution of expression and the number of genes that analyst intend to analyze at one point of time. In this, usually crude RNA/cDNA is used as a source material. Expression in tissue section or cells can be assessed by tissue/cell in situ hybridization. Expression analysis in isolated RNA can be done by northern blot hybridization, RNA dot-blot hybridization, or real-time PCR/qPCR. These are low-through put expression screens which analyze expression of only one gene or a very small number of genes at a time. In this chapter, the readers would learn about isolation of total RNA from peripheral blood mononuclear cells, agarose gel electrophoresis for quality check of RNA, cDNA synthesis, and real-time PCR reaction. These experiments can be undertaken when the researcher needs to study gene expression in disease versus control or in response to stress/stimuli.

Keywords PCR, Real-time PCR, C_T values, End-point analysis

1 Introduction

Polymerase chain reaction (PCR) is an extremely simple, user friendly, and popular technique [1]. It has many applications that are dependent on its features: (1) swift—the technique results in quick amplification of target DNA, (2) sensitive—it is highly sensitive, it is capable of amplifying very minute quantity of DNA, a

Neelima Gupta and Varsha Gupta (eds.), *Experimental Protocols in Biotechnology*, Springer Protocols Handbooks, https://doi.org/10.1007/978-1-0716-0607-0_12,

single cell can be typed using the technology, and (3) robust—it is highly robust as badly damaged or degraded DNA or complexed DNA or decomposed DNA from historical sites can be amplified using PCR [2].

As the PCR reaction is based upon annealing of primer at the specific site thus it is capable of amplifying a specific target in heterogeneous DNA sequences.

The reaction consists of three cycles which are based upon temperature, that is, denaturation, annealing, and extension. Denaturation is often achieved at 92–95 °C, temperature is then lowered to allow base pairing of the primer (synthetic oligonucleotides) with target sequence called annealing and finally wherever primer binds, the DNA polymerase at extension temperature utilizes the components of PCR reaction mix as dATP, dGTP, dTTP, and dCTP along with buffer salts to extend these primers using source DNA as template (extension). Thus, the process is also known as thermocycling and the machine is referred to as a thermocycler [1].

The reaction is exponential, that is, orientation of primer is such that the direction of new strand created is toward the primer binding site. Ideally, the thermocycler sequentially repeats denaturation, annealing, and DNA synthesis or extension for about 30–40 times. The temperature also changes in the reaction; therefore, it is essential that our DNA polymerase enzyme used in the reaction should be able to withstand such high temperatures. Taq DNA polymerase is the popular enzyme obtained from *Thermus aquaticus.* This bacterium resides in natural hot springs. However, Taq DNA polymerase lacks $3' \rightarrow 5'$ exonuclease activity that provides the proofreading ability to the enzyme, making it prone for incorporating occasional errors. Therefore, extreme thermophilic archaeon *Pyrococcus furiosus* was sought, and it is a source of heat-stable DNA polymerase with $3' \rightarrow 5'$ proofreading exonuclease activity; the enzyme is referred to as Pfu polymerase. The template can be genomic DNA or it can be cDNA for transcript analysis [2].

2 Technique

Quantitative PCR or real-time PCR is emerging as a powerful tool for quantitation of DNA/RNA. Importantly, for gene expression analysis, RNA is used as starting sample. RNA (transcribed part of genome) from suitable tissues can be isolated and using it as the template, complementary DNA (cDNA) can be produced by using reverse transcriptase enzyme [2, 3]. This reaction is known as reverse-transcriptase PCR or RT-PCR. Thus, if the RNA transcript of the target gene is low in a particular sample, the resultant signal would be low in contrast to high RNA transcript of the target gene in another sample.

Normally, the PCR reactions are based on end point analysis, that is, the amplified products are analyzed at the end of the complete reaction by size fractionation on agarose gel electrophoresis. These gels are stained by ethidium bromide which fluoresce under ultraviolet light. Thus, these reactions are not very well suited for quantitative determination. In real-time PCR reaction, real-time monitoring can be done when the reaction is actually occurring in specialized equipment. The reaction products can be monitored at early stage of amplification during exponential cycle which allows for precise quantitation using fluorescent probes.

3 Materials

The materials required for different experiments are listed below:

3.1 Peripheral Blood Mononuclear Cell Isolation

Blood sample (PBMCs), ice-cold phosphate buffer saline (PBS—137 mM NaCl, 2.7 mM KCl, 10 mM Na_2HPO_4, 2 mM KH_2PO_4 pH-7.4 adjusted by 5 M HCl), TRIzol, chloroform, microcentrifuge tube, isopropanol, 70% ethanol, RNase-free water or DEPC-treated water, tips 10 μL, 200 μL and 1000 μL, gloves, refrigerated centrifuge, micropipettes 2–20 μL, 2–200 μL, and 100–1000 μL, vortex or homogenizer.

3.2 Agarose Gel Electrophoresis

1.5 to 2.0% agarose, 1× Tris–acetate–EDTA buffer (TAE buffer—40 mM Tris base, 1 mM EDTA [pH 8.0], 1.142 mL glacial acetic acid [1 l]), 1% EtBr (ethidium bromide), electrophoretic unit, power supply, casting apparatus with comb, micropipettes, microtips.

3.3 cDNA Synthesis

RevertAid first strand cDNA synthesis kit, microcentrifuge tubes (200–500 μL), RNase-free water, tips 0.1–2.5 μL and 2–200 μL, centrifuge, temperature-regulated water bath or dry bath, micropipettes.

3.4 Real-Time PCR Reaction

cDNA as template, microtips 0.1–10 μL and 2.0–200 μL, qPCR tubes 0.2 mL, 8-well strip, nuclease-free water, target gene primer (forward and reverse—Stock 100 pM), SYBR Green qPCR master mix, centrifuge, micropipettes, real-time PCR machine.

4 Methods

4.1 Total RNA Isolation by Using Acid Guanidinium Thiocyanate–Phenol–Chloroform (TRIzol)

4.1.1 Principle

RNA isolation by using acid guanidinium thiocyanate–phenol–chloroform (TRIzol) is useful and gives good results when samples are rich in endogenous RNases. TRIzol is a solution consisting of guanidinium isothiocyanate and phenol which deproteinizes RNA [4]. It disrupts and breaks the cells and denatures the protein. It maintains RNA integrity while extraction with chloroform separates aqueous and organic phases. RNA is retained in the aqueous phase and recovered through centrifugation by isopropanol precipitation method.

4.1.2 Procedure

Use 300–500 μL blood (3×10^5 PBMCs) and add equal volume of PBS. Wash the solution by centrifugation at 3000–5000 rpm for 5 min at 4 °C. To the pellet, add 300–500 μL TRIzol reagent and vortex or homogenize properly with homogenizer or by repeated pipetting. To this, add 60–100 μL chloroform (20% of chloroform per mL TRIzol) and incubate for 10 min at room temperature (RT). Centrifuge at 10,000 rpm for 10 min at 4 °C. Three layers are observed, upper aqueous, middle white and lower pink due to the color of TRIzol. Separate the upper aqueous phase in a new fresh microcentrifuge tube and repeat the process from chloroform addition. Again, three layers appear, collect upper aqueous phase in fresh microcentrifuge tube. Now add equal volume (1:1) of isopropyl alcohol and incubate for 10 min at room temperature (RT) and centrifuge at 10,000 rpm for 10 min at 4 °C. Add ice-cold 70% ethanol and wash properly by centrifugation for 10 min at 10,000 rpm. Remove the supernatant and allow the pellet to dry for 5 min. Add 30–50 μL RNase-free or DEPC-treated water to dissolve the pellet and store at −80 °C in aliquots.

4.1.3 Expected Outcomes

After washing with 70% ethanol, a small pellet is observed at the bottom of the centrifuge tube. This total RNA extracted from PBMCs is used to prepare cDNA using reverse transcriptase enzyme to determine the level of expression of various genes in our samples. With this methodology, RNA can be easily isolated. RNA isolation from PBMCs is a good model to study the efficacy of new drugs by gene expression analysis.

4.1.4 Precautions

1. All the glassware should be properly cleaned and baked at 200 °C for 24 h.
2. All the plasticware should be properly washed, treated with DEPC water and autoclaved.
3. The working room should be closed and disinfected with disinfectants.

4. The integrity and purity of total RNA is an important factor in the synthesis of high-quality cDNA.
5. Total RNA should be free from DNA contamination.
6. Gloves should be worn throughout the procedure to protect the RNA samples from degradation by nucleases.
7. RNase-free or DEPC-treated water should be used in preparing the buffers.

4.2 Agarose Gel Electrophoresis for Separation of RNA

4.2.1 Principle

Agarose gel electrophoresis is used in several fields as biochemistry, molecular biology, and immunology to separate and visualize DNA, RNA, and proteins. Agarose is used as the gelling material through which molecules migrate under the influence of electric field and separate on the basis of their charge–mass ratio. Agarose is a polysaccharide obtained from red sea weeds. It is a linear molecule consisting of repeating units of agarobiose (disaccharide formed by D-galactose and 3,6-anhydro-L-galactopyranose) [5].

Mass and molecular conformation of biomolecule are important in determining the speed of their movement in the gel as both DNA/RNA have net negative charge on them [6]. Due to negative charge, under electric field, both the biomolecules migrate towards anode on the basis of their mass/size and conformation [7] which is ultimately affected by frictional forces active during migration. More is the friction, lesser is the mobility. By analyzing intensity of band and its position, information about quantity and quality of the biomolecule can be established. The separated nucleic acids can be stained by ethidium bromide and visualized under UV light. Most Agarose gels are casted with 0.6–2% agarose concentration depending on size of target molecules.

After isolation of RNA, it is very important to analyze the quality and quantity of RNA. Quantity is analyzed by taking absorbance at 260 nm (nucleic acids) and 280 nm (protein-aromatic amino acids as tyrosine, tryptophan and phenylalanine absorb at this wavelength). Ideally, the ratio of 260/280 should be close to 1.8–2.0. However, this absorbance does not give true indication about DNA contamination or damaged RNA/DNA Fig. 1. DNA contamination and degraded RNA both add to absorbance at 260 nm. Thus, quality can be analysed by running agarose gel electrophoresis and visualization of bands after staining with ethidium bromide. Due to its hydrophobic nature, ethidium bromide is able to insert itself and interacts through Vander Waals contact with stacked bases in DNA/RNA. It is thus referred to as intercalating agent. Under ultraviolet light, it fluoresces, emitting an orange color [8].

Agarose gel electrophoresis gives an estimate about purity and quality of RNA. In agarose gel electrophoresis, the charge molecules like DNA and RNA migrate towards their oppositely charged electrode under the influence of electric field. The rate of migration

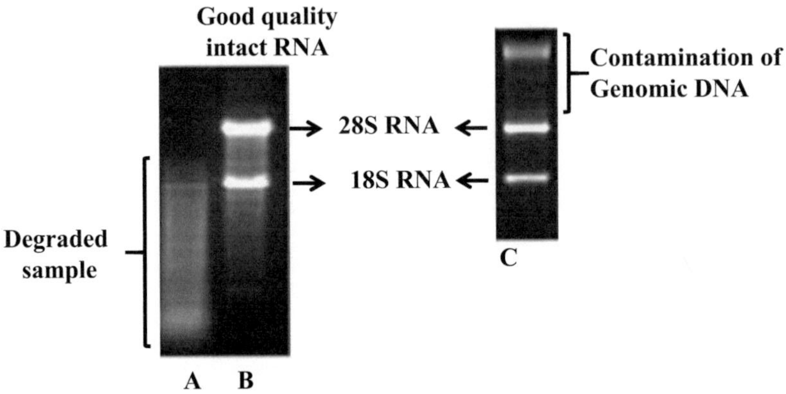

Fig. 1 Degraded sample (**a**), good quality intact RNA (**b**), and good quality RNA with DNA contamination (**c**)

of DNA/RNA depends on its molecular weight, conformation and porosity of gel.

4.2.2 Procedure

Dissolve 1.5–2.0% (w/v) agarose in 1× TAE buffer (tris–acetate–EDTA) buffer and heat in a microwave oven until the agarose dissolves. Measure the volume and maintain it with buffer. Allow it to cool, till temperature reaches 45–50 °C and add 1–2 μL ethidium bromide (stock 0.5 μg/mL) and mix well. Pour in a sealed gel casting tray with a comb. The gel is poured into the apparatus and allowed to solidify. It is immersed into the electrophoretic tank buffer and the comb is carefully removed.

After recording absorbance (260 nm and 280 nm), equal quantities of RNA sample with 2× RNA gel loading dye is loaded into the wells with the help of micropipette avoiding bubbles. Standard molecular weight ladder is always loaded along with the samples, as it helps in assessing the size of the RNA fragment.

After completion of run, place the gel under gel documentation system or UV transilluminator to visualize the RNA bands. The ethidium bromide intercalated in the RNA fragments emit orange fluorescence under UV light.

4.2.3 Expected Outcome

This experiment gives an estimate about the quality and cross contamination in the sample. Best results are obtained with good quality non contaminated samples.

4.2.4 Precautions

1. The gel apparatus, comb and tray should be properly treated with DEPC water.
2. Ethidium bromide is a mutagen, therefore careful handling and disposal is required.
3. RNA is sensitive and prone to degradation, thus requires careful handling.

4.3 cDNA Synthesis by Reverse Transcriptase Enzyme

4.3.1 Principle

Formation of DNA using information from RNA template by RNA dependent DNA polymerase is reverse transcription. RNA-directed DNA polymerases are referred as Reverse transcriptase (RTs) which were discovered in viral life cycle. Using short complementary primer, reverse transcriptase synthesizes first strand of complementary copy of RNA which is termed as cDNA or complementary DNA. This cDNA can directly be used as template for PCR reaction. The technique of reverse transcription allow the detection of RNA with lower abundance.

cDNA synthesis allow cloning and detection of low copy number transcripts. cDNA can also be made double stranded using DNA polymerase I and DNA ligase. Common reverse transcriptase (RT) are either Avian Myeloblastosis Virus (AMV) RT or Moloney Murine Leukemia Virus (M-MuLV, MMLV) RT. RT catalyzes the synthesis of DNA by using RNA as the template. Like other DNA polymerases they also extend the primers provided in the reaction. The reaction requires buffer salts, dNTPs for efficient reaction. There are two kinds of natural enzymes. One of the reverse transcriptases is Avian Myeloblastosis Virus (AMV) reverse transcriptase (RT) which requires Mg^{2+} or Mn^{2+} for activity. It is a thermostable enzyme which decreases hairpin and secondary structure formation, thus less sensitive to inhibition by strong RNA secondary structures. AMV RT is compatible with gene-specific primers, oligo (dT)15 primers, or random hexamers.

RT from Moloney Murine Leukemia Virus (M-MLV) is used for reverse transcription of lengthy RNAs (>5 kb). It possesses lower RNase H activity and needs lesser reaction temperatures when compared to AMV RT. Moloney-MLV RT is prone for inhibition by RNA secondary structure formation. Now mutant forms are available that have bypassed all these limitations of wild-type M-MLV RT.

In several detection methods, when transcripts are detected, it is extremely difficult to detect low copy number transcripts. Therefore, PCR based methods are more sensitive and are able to track expression of genes in different cell types or tissues that are otherwise difficult to analyze. There are several methods for the purpose but almost all of them require the copying of starting RNA into cDNA sequence by using RT.

The basic technique involves utilizing poly-A tail of mRNA. Oligo (dT) primer or random sequence oligonucleotide primers are used to initiate a PCR reaction. As it is difficult to quantitate cDNA therefore samples are equalized on the basis of starting RNA concentration.

In this section, the readers would learn about the methodology of cDNA formation from mRNA by RT-PCR reaction.

4.3.2 Procedure

Synthesis of cDNA is basically done from total RNA. The mRNA has a poly-A tail which helps in the binding of oligo$(dT)_{18}$ primers for the extension process of cDNA synthesis. Oligo (dT) are short length of thymine nucleotides that binds at 3′-OH end of poly-A tail of the mRNA. The reverse transcriptase enzyme uses mRNA as the template and synthesizes the cDNA. The initial step of DNase I treatment before cDNA synthesis step is done.

DNase I Treatment

Isolated total RNA should be free of DNA contamination. For removal of DNA contamination, treat RNA with DNase I before starting the synthesis of cDNA.

1. In RNase-free tube, add 0.1–5 μg total RNA as templates.
2. Add 1 μL (10× with $MgCl_2$) reaction buffer.
3. Add 1 μL DNase I (1u).
4. Finally, add nuclease-free water to makeup final volume up to 10 μL.
5. Incubate the tubes at 37 °C for 30 min.

Briefly centrifuge so that the components sticking at the walls settle at the bottom to initiate the synthesis of cDNA from RNA samples. DNase can be removed by heating or phenol-chloroform extraction before processing for cDNA synthesis.

cDNA Synthesis

For cDNA synthesis follow the instruction manual

1. Use 0.1 ng to 5 μg RNA samples in 200–500 μL microcentrifuge tubes.
2. Add 1 μL oligo $(dT)_{18}$ primer and water to make up 12 μL volume.
3. Add 4 μL 5× Reaction Buffer, 1 μL RiboLock™ RNase inhibitor (20 u/μL), 2 μL dNTP Mix (10 mM), and finally 1 μL RevertAid™ M-MuLV reverse transcriptase (200 u/μL) in the reaction mixture to make 20 μL final volume.
4. Gently mix the components and spin down by centrifugation and incubate at 42 °C for 60 min.
5. Terminate the reaction by heating at 70 °C for 5 min.

4.3.3 Expected Outcome

Synthesis of first strand cDNA from mRNA in total RNA population is obtained from source. cDNA can be utilized for gene expression analysis. Prepared cDNA can be used for multiple applications as DNA microarray analysis, cDNA library preparation or gene expression analysis using semiquantitative or quantitative PCR using real-time PCR machine.

4.3.4 Precautions

1. All the laboratory wares should be properly treated.
2. As quantitation of cDNA is difficult, RNA should be properly quantified by measuring absorbance and its quality check should be done using agarose gel electrophoresis.
3. cDNA may be assumed to be equivalent to total RNA concentration used in the reaction as it is first strand cDNA synthesis.
4. The cDNA can be analyzed by ethidium bromide staining after subjecting it for agarose gel electrophoresis, it has several limitations as the sensitivity of detection becomes very low and accurate quantification is difficult.

4.4 Data Evaluation of Gene Expression by Quantitative PCR (qPCR)

Real-time PCR reaction can be used for absolute or relative quantification of the sample. Absolute quantification is used when we are interested in analyzing properties which are intrinsic to the sample as the copy number of the target.

Relative quantification is used when we are interested in analyzing differences in gene expression in response to treatment, diseases, developmental processes, and so on. It is used to analyze fold changes in gene expression. The present chapter will deal with relative quantification of the sample.

4.4.1 Principle

In this method unlike the basic PCR method, real-time PCR (qPCR) does not involve random primers and analysis by gel fractionation. In this reaction, specific primers are used for the target gene.

The technique of qPCR is a kinetic reaction which requires specialized PCR equipment. In this qPCR, the amplification products are continuously quantitated during the PCR reaction. For obtaining the accurate data, the estimation of quantitation is done at early stages of PCR reaction, when the amplification is still exponential. As detection of amplification products during a qPCR reaction requires some measurable signal, therefore fluorescence compounds as probes are utilized. The fluorescent signal is proportional to the amount of the amplified product.

In this, detection of specific gene is done by using primer which binds to specific template in the qPCR reaction. This uses a fluorophore at the 5′ end. The fluorophore, SYBR Green dye shows little fluorescence when it is free in solution. After binding to double stranded DNA, the fluorescence of the dye is increased to more than 1000-fold. But sometimes, it can also bind to primer-dimers or nonspecific amplification products. To control this, melting curve analysis is done at the end by increasing the temperature slowly from 60 °C to 95 °C with continuous monitoring of the fluorescence. At a certain temperature, whole of amplified product will fully dissociate, leading to decrease in fluorescence, as the dye dissociates from the DNA. This dissociation temperature is dependent on the composition and length of the amplicon, allowing different DNA fragments to be distinguished [1].

Thus, the real-time PCR is similar to conventional or standard PCR, but the product is monitored during real-time interval with the help of fluorescent dyes or probes. There are various dye or probes used to detect the amplification (SYBR Green I, TaqMan probes, molecular beacon probes, etc.). SYBR Green dye does not bind with single stranded DNA (ssDNA) and gives very small fluorescence in the solutions while during PCR amplification it binds with dsDNA and gives a strong fluorescence that can be measured by detectors and signals can be seen on the computer screen.

4.4.2 Procedure

Set up the real-time reaction in 20 μL with cDNA (20 pM), SYBR Green master mix, 1 μL primer pair, and water for volume makeup. The protocol with specific annealing conditions is required to be carried out in the PCR machine with continuous monitoring.

Reaction master mixes includes dNTPs, Taq DNA polymerase, $MgCl_2$, and SYBR Green I dye (a fluorescent dye) or probes (TaqMan® probe).

Component	Volume per reaction (μL)
Supermix (2×)	5.000
10 pM forward primer (SYBR_target_1)	0.200
10 pM reverse primer (SYBR_target_1)	0.200
Template	1.00
Water	3.600
Total volume (excluding template)	9.000

qPCR Protocol

1. Initial denaturation: 95.0 °C for 1:00.
2. Denaturation: 95.0 °C for 0:05.
3. Annealing: 58.0 °C for 0:30 (specific for primer).
4. Plate Read.
5. GOTO 2, 39 more times.
6. Melt Curve 60.0 °C to 95.0 °C: Increment 0.5 °C 0:05.
7. Plate Read.

4.4.3 Expected Outcomes

The analyst can see the fluorescent signal in real time rather than going for end point analysis as in traditional PCR reaction (Fig. 2). Real-time PCR being the kinetic reaction provides the user an opportunity to monitor the levels of signal during early stages of PCR reaction using fluorescent probe.

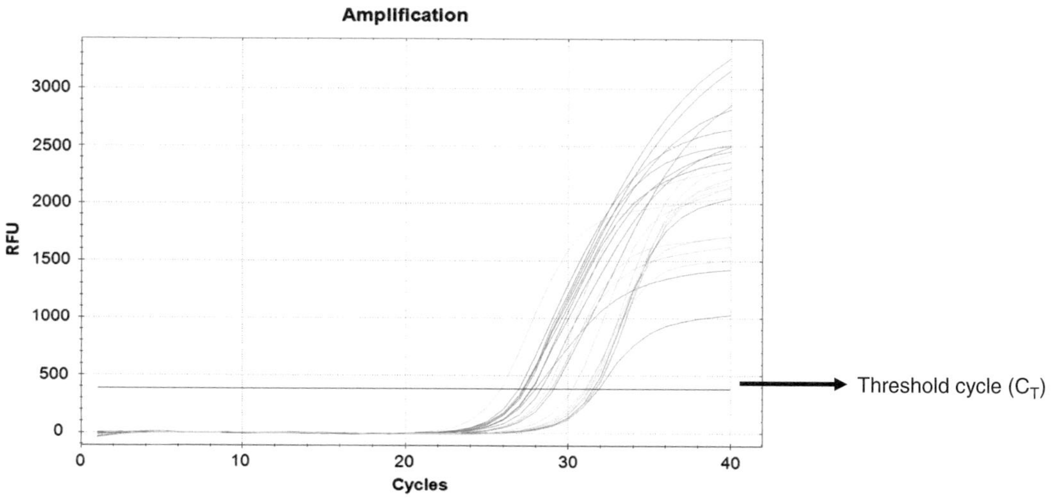

Fig. 2 Fluorescence from amplification reaction on *Y*-axis and number of PCR cycles on *X*-axis. ∗During early cycles, fluorescence is at background levels and thus is not quantified. ∗With time, fluorescence increases as sufficient amplified products are generated. The products are detectable at C_T or threshold cycle

C_T Value

The C_T values in real-time PCR is the required cycles to reach a threshold so as to emit fluorescence signals and cross the background levels (Fig. 2). Threshold cycle (C_T) is dependent upon the amount of template present in the reaction which, in turn, is dependent upon the amount of our starting material (in this case, starting mRNA transcripts).

For example, if in our sample large amount of RNA transcript is available for the target gene that means there would be high amount of target cDNA template. Thus, C_T (for detectable signal) would be lower as only few cycles would be needed for sufficient buildup of the product to achieve detection above background. In the other case, if our target RNA transcript is low in the sample, that would lead to the corresponding lower formation of cDNA, therefore, as the template is low, more amplification cycles would be required for detectable signal resulting in higher C_T value.

Higher C_T—Small amount of template is present in the sample.

Lower C_T—Large amount of template is present in the sample.

Thus, high transcript concentration allows the fluorescence signal to cross the threshold quickly; thus, C_T value is inversely proportional to the amount of target genes present into the reaction volume. Whereas lower starting RNA gives higher C_T, higher target RNA yields lesser C_T values.

C_T values, if less than or equals to 29 cycle is the sign of strong positive reaction in which abundant amount of target templates is present in the sample.

C_T values, if between the 30 and 35 cycles, it shows the sign of moderate positive reaction in which moderate amount of target templates is present in the sample.

C_T values, if greater than or equals to 35–40 cycle is the sign of weak positive reaction in which small amount of target templates is present in the sample.

Normalized Expression Level Calculation

As we are working on relative quantification therefore, we are interested in observing fold changes in gene expression. Relative quantification requires a calibrator or control. Control ensures that the comparison is made between equal quantity of starting sample. There can be two methods for normalization [2] which are illustrated below:

1. Normalization against a reference unit of mass [2].
2. Normalization against a reference gene [2].

<table>
<tr><th>Normalization against a reference unit of mass</th><th colspan="3">Normalization against a reference gene</th></tr>
<tr><td>Requires accurate quantification of the transcript</td><td colspan="3">If accurate quantification is not available or starting material is limited</td></tr>
<tr><td>One sample is calibrator or control; the other is test sample, expression is expressed as increase or decrease relative to the control</td><td colspan="3">One sample is control and the other is test; however, it necessitates the use of known reference gene which shows stable expression between the groups compared</td></tr>
<tr><td>C_T values of control and test are used for calculating ratio of test/control</td><td colspan="3">C_T values of target gene and reference genes are required in healthy control and disease</td></tr>
<tr><td rowspan="5">$\text{Ratio(test/control)} = E^{C_T(\text{Control}) - C_T(\text{Test})}$
Here E is reaction efficiency, assuming that the reaction is optimized with very good efficiency. It indicates that every cycle result in exact doubling of the template; thus, the equation can be expressed as
$\text{Ratio(test/control)} = 2^{C_T(\text{Control}) - C_T(\text{Test})} = 2\Delta^{C_T}$
where $\Delta C_T = C_T\ (\text{Control}) - C_T\ (\text{Test})$</td><td>Sample</td><td colspan="2">Gene</td></tr>
<tr><td></td><td>Analyzed target gene</td><td>Reference Gene</td></tr>
<tr><td>Control</td><td>C_T (target gene, control)</td><td>C_T (reference gene, control)</td></tr>
<tr><td>Disease</td><td>C_T (target gene, disease)</td><td>C_T (reference gene, target)</td></tr>
<tr><td colspan="3">It requires 1 reference gene (e.g., GAPDH) and one target gene (e.g., IL8)</td></tr>
<tr><td>Example: If we have observed C_T values in our rheumatoid arthritis (RA) disease and healthy control group for gene IL-1 as
Control $C_T = 15$
Test (RA) $C_T = 12$
Calculating
$\text{Ratio(test/control)} = 2^{C_T(\text{Control}) - C_T(\text{Test})}$
$= 2^{(15-12)}$
$= 8$
Therefore, expression of this gene is eightfold high in RA when compared to healthy control</td><td colspan="3">Livak method or $\Delta\Delta C_T$
1. Requires two values for ΔC_T
2. ΔC_T of the analyzed target gene is normalized to ΔC_T of reference in control
3. ΔC_T of the analyzed target gene is normalized to ΔC_T of reference in disease
4. The expression ratio of fold difference is calculated</td></tr>
</table>

Using Livak method or $\Delta\Delta C_T$ method to relative quantification [2]:

Let us compute this example—The C_T values for gene expression of IL-8 and glyceraldehyde 3-phosphate dehydrogenase (GAPDH) are shown.

	Gene	
Sample	C_T (Analyzed target gene) (IL8)	C_T (reference gene) (GAPDH)
Control	15	16
Disease	12	16
It requires 1 reference gene (e.g., GAPDH) and one target gene (e.g., IL8)		

In the first step C_T of the analyzed target gene is normalized to the C_T of the reference gene.

1. ΔC_T (control) = C_T (control, IL8) − C_T (control, GAPDH)
   ΔC_T (control) = 15–16 = −1
2. ΔC_T (disease) = C_T (disease, IL8) − C_T (disease, GAPDH)
   ΔC_T (disease) = 12–16 = −4 (both control and disease have given negative C_T values).
3. ΔC_T (control) = 15–16 = −1
   ΔC_T (disease) = 12–16 = −4
4. ΔC_T of the disease is normalized by ΔC_T of the control
   $\Delta\Delta C_T = \Delta C_T$ (disease) − ΔC_T (control)
   $= -4 - (-1) = -3$
   Thus $\Delta\Delta C_T = -3$
   Determining the fold difference in the expression of target gene IL8 in disease and control is calculated as $= 2^{(-\Delta\Delta C_T)} = 2^{-\Delta\Delta C_T} = 2^{-(-3)} = 8$. Therefore, disease subject in question express this gene at eight-fold higher level than control. When negative values are obtained, they show that difference is lower in disease/test than the control.
   $\Delta\Delta C_T$ is also denoted by C_q and referred as comparative quantification methodology analysis by qPCR. The formula calculates the fold changes of the target gene expression levels by the qPCR.

5 Notes

1. During RNA synthesis, RNA should be treated with DNase I to remove the genomic DNA contaminations [9].
2. There should be no genomic DNA contamination in the cDNA synthesized from RNA [9].

3. Primers need to be properly designed and evaluated for primer-dimer formation [10].
4. Accurate quantification of starting RNA is very important as it is extremely difficult to quantify cDNA [11].
5. Accurate pipetting and proper precautions should be taken as minor error may lead to big changes in the results.
6. Gloves should be worn during all the procedures.

Troubleshooting:

(a) DNA contamination in RNA

There are traces of DNA that contaminate RNA, thus proper phenol, chloroform, isoamyl treatment can tackle this issue. Very careful removal of aqueous phase can help.

Before processing for cDNA synthesis DNase I treatment would be helpful. Proper use of TRIzol buffer and careful removal of upper aqueous phase along with DNase I treatment usually solve these problems. Genomic DNA contamination leads to unwanted signals [9].

(b) Degraded RNA or low-integrity RNA

RNA is a very sensitive biomolecule; thus, its isolation requires special care and proper clean environment. All glassware and plasticware should be carefully treated. RNA after isolation should be aliquoted, its quantitation is recommended before processing for RT reaction. Adding 4-mercaptoethanol in lysis buffer may also help; however, guanidinium salts–based buffers are efficient in preventing RNA degradation. Frozen samples require more care for RNA isolation. All the buffers should be carefully prepared for suspending RNA. RNA should be aliquoted and once one aliquot is used, its reuse should be avoided. Always use nuclease-free water for all the steps [9].

(c) RNA yield is low

Check for sample that you have processed. Whether it is fresh, stored, freeze-thawed sample, quality of sample is important for good-quality RNA.

(d) Reverse transcription of RNA is not efficient

Ensure complete removal of guanidinium salts and phenol. These not only inactivate RNases but also inhibit other enzymes. Proper washing of sample for removal of these inhibitors helps and gives better outcome. Use efficient and thermostable reverse transcriptase because at high temperatures secondary structures of RNA are prevented. Sometimes hairpins loops and secondary structures are formed, thus, performing RT reaction at elevated temperatures may help.

(e) Check for quality of RNA

It is very important that RNA quality should be checked using agarose gel electrophoresis.

(f) Nonspecific amplifications in Real-time PCR reaction

Think of redesigning the primer pairs and check for genomic DNA contamination. DNase I treatment helps; try to perform reverse transcription as well as actual real-time cycles at high temperatures to avoid nonspecific amplification. If cDNA sample is contaminated, it will tend to give nonspecific results.

References

1. Strachan T, Read A (2012) Human molecular genetics, 4th edn. Garland Science group (Routledge Taylor and Francis group), New York; ISBN: 9780815341499
2. Bio-Rad Real Time PCR manual. https://www.bio-rad.com/webroot/web/pdf/lsr/literature/
3. Chomczynski P, Sacchi N (1987) Single-step method of RNA isolation by acid guanidinium thiocyanate-phenol-chloroform extraction. Anal Biochem 162(1):156–159
4. Jeppson JO, Laurell CB, Franzen B (1979) Agarose gel electrophoresis. Clinical Chem 25 (4):629–638
5. Kryndushkin DS, Alexandrov IM, Ter-Avanesyan MD, Kushnirov VV (2003) Yeast [PSI+] prion aggregates are formed by small Sup35 polymers fragmented by Hsp104. J Biol Chem 278(49):49636–49643
6. Sambrook J, Russel DW (2001) Molecular cloning: a laboratory manual, 3rd edn. Cold Spring Harbor Laboratory Press, New York
7. Moran L, Mirault M-E, Tissières A, Lis J, Schedl P, Artavanis-Tsakonas S, Gehring W (1979) Physical map of two *D. melanogaster* DNA segments containing sequences coding for the 70,000 Dalton heat shock protein. Cell 17:1–8
8. Dheda K, Huggett JF, Chang JS, Kim LU, Bustin SA, Johnson MA, Rook GA, Zumla A (2005) The implications of using an inappropriate reference gene for real-time reverse transcription PCR data normalization. Anal Biochem 344(1):141–143
9. https://lifescience.roche.com/en_in/articles/dna-free-rna-isolation-rna-isolation.html
10. Brownie J (1997) The elimination of primer-dimer accumulation in PCR. Nucleic Acids Res 25(16):3235–3241
11. https://www.bio-rad.com/en-in/applications-technologies/real-time-pcr-troubleshooting?ID=LUSOBDHYP

Chapter 13

Use of Droplet PCR in Biomedical Research

Kyle A. Doxtater, Manish K. Tripathi, Murali M. Yallapu, Meena Jaggi, and Subhash C. Chauhan

Abstract

With the development of semiquantitative PCR, scientists were able to begin looking at specific genes; however, this technique had several limitations. Specifically, semiquantitative PCR is an end point study, only yielding information on the expression of the target gene at the end of several amplification cycles, which, in turn, limits its ability to quantify gene expression based only on signal intensity. This signal could be affected by various factors, such as the number of cycles performed, and often the final product is oversaturated. In order to combat this limitation, scientists developed real-time PCR (qRT-PCR). qRT-PCR allowed for a "real time" view of gene expression by taking a reading after each individual amplification cycle. With these successive readings, qRT-PCR allows for a relative quantification of gene expression between an experimental group and its control using the "double delta" (2Δ equation). However, quantitative PCR has its own limitations. To address these limitations, a process known as digital droplet PCR (ddPCR) was developed. In this chapter readers would learn about the ddPCR technique and its applications.

Keywords PCR, Digital droplet PCR, End point analysis

1 Introduction

Digital PCR is a powerful technique that finds utility in many applications. The technology is sensitive for quantification of nucleic acids in the sample. A robust technique capable of detecting targets with low abundance and complicated targets with high background. Digital PCR provides sensitive and absolute quantification with precision. It is capable of detecting.

1. Variations in copy number: Differences in the copies of the particular target in control and test subjects. It may or may not affect phenotype but leads to genetic variability.
2. Single nucleotide polymorphism (SNP) detection: It is capable of detecting SNP and rare sequences in complicated background.

Neelima Gupta and Varsha Gupta (eds.), *Experimental Protocols in Biotechnology*, Springer Protocols Handbooks, https://doi.org/10.1007/978-1-0716-0607-0_13,

3. Pathogen detection: The technology along with other powerful tools as next-generation sequencing is enabling sensitive and absolute estimation of microorganisms, their identification etc.
4. Microbiome analysis: It is capable of quickly detecting pathogens, their strains along with drug resistance information, which makes it a versatile tool for microbiology research.
5. Quantifying pathogenicity and pathogen load: The technology is capable of detecting infectious agent and one can also detect viral load and thus associated viremia after therapy.
6. Gene expression and its epigenetic regulation: It is capable to quantify expression of rare genes and expression analysis with high precision. It is also capable of detecting epigenetic events controlling expression of genes in the cell.

2 Technique

Polymerase chain reaction (PCR) has been a significant building block in the progression of biological sciences since its discovery in the 1980s [1]. After that, amplification of DNA became a routine exercise in the laboratories all over the world. With smart technologies, quantitative PCR or real-time PCR or qRT-PCR was invented which bypassed end point analysis, making analysis of gene expression easier. However, both the techniques had their own limitations. Like semiquantitative PCR, qRT-PCR has its own limitations, often not being sensitive enough to detect rare events and requires significant amount of material in comparison to other PCR techniques [2–4]. Digital droplet PCR (ddPCR) was first developed in 2011. This technique improves quantitative accuracy by partitioning the sample into thousands of individual droplets, which, in turn, allows for thousands of discrete reactions per sample, leading the way to absolute quantification with high precision and sensitivity. This overcomes the limitations posed previously in qRT-PCR. ddPCR due to its high sensitivity requires as little as 0.01 ng of input nucleic acid per assay for a signal. ddPCR has been shown in literature to have applications in liquid biopsy, rare mutation detection (RMD), next-generation sequencing (NGS), single cell analysis, gene editing event detection, gene expression and miRNA analysis, and residual DNA detection [3–16].

3 Materials

QX200 ddPCR EvaGreen Supermix (Bio-Rad Cat#1864034).

QX200 Droplet Generation Oil for EvaGreen (Bio-Rad Cat#1864006).

QX200 Buffer for EvaGreen (Bio-Rad Cat#1864052)

ddPCR Supermix for probes (no dUTP) (Bio-Rad Cat#1863024)

Droplet Generation Oil for probes (Bio-Rad Cat#1863024) ddPCR Buffer Control for probes (Bio-Rad Cat#1863052) ddPCR Droplet Oil (Bio-Rad Cat#1863004)

DG32 Automated Droplet Generator Cartridges (Bio-Rad Cat#1864108)

ddPCR 96-Well Plates (Bio-Rad Cat#12001925)

PCR Plate Heat Seal, Foil, Pierceable (Bio-Rad Cat#1814040).

4 Methods

4.1 Stepwise General Overview of ddPCR

4.1.1 Step 1: Sample Preparation

Samples are prepared according to experimental needs. RNA, complementary FDNA (cDNA), or genomic DNA (gDNA) is mixed with the appropriate Supermix. This Supermix is specific to each PCR reaction and can be either TaqMan (hydrolysis) or EvaGreen (SYBR Green) depending on experimental needs.

4.1.2 Step 2: Cycling of Droplets

Droplets for each sample are then generated by drawing a nanoliter-sized portion of the sample and incorporating it into a droplet of oil to ensure uniformity. Droplets are then run through and end point PCR reaction using a deep well cycler (C1000 Thermal Cycler) to ensure even heating throughout the well. The ramp rate is set to 2°C/s to create a more evenly heated sample and is then held at 12 °C before reading in the QX200.

4.1.3 Step 3: Reading Droplets

The experiment is set up in QuantaSoft Analysis Software designating the well name, experiment type, and designation of assay corresponding to fluorescent channels. The four main experiment types are direct quantification, copy number variation, drop off, and gene expression. The software will look for a one-channel (single-plex) assay or two-channel (duplex assay) and the resultant readout will be based on the experiment type selected. Refer to Digital Droplet PCR Applications Guide (http://www.bio-rad.com/webroot/web/pdf/lsr/literature/Bulletin_6407) for additional instructions on analysis methods.

4.2 ddPCR Protocol

1. Harvest cells to generate relevant gDNA or RNA based on experimental design.

Table 1
Components and their quantity for reaction mix

Reagent	Volume	Master Mix
Supermix	12.5 μL	Yes
Primer/probes	0.5–1 μL	Yes
Water	(10–9.5 μL) – *X*	Yes
Sample	*X*	No

Note that for cDNA there should be no more than 350 ng/well and no less than 0.01 ng/well. For genomic DNA, there should be less than 66 ng without restriction enzymes or greater than 1 μg with restriction enzymes. Add sample to Master Mix and vortex.

2. Convert harvested cells into proper Nucleic Acid format for experiment (whether creating cDNA, RNA, or gDNA).
3. Prepare PCR reaction sample -final volume will be 22–25 μL per well. Make sure to use appropriate Supermix for PCR reaction (TaqMan or EvaGreen). Supermix must be at least 50% of the final volume. Use Table 1 to create reactions.
4. Add DGB cartridge to DGB cartridge holder.
5. Add 20 μL of sample to sample row of DGB cartridge.
6. Add 70 μL of droplet generation oil to the oil row of DGB cartridge.
7. Attach gasket to DGB cartridge holder making sure holes align properly.
8. Place setup in droplet generation machine.
9. Add 40 μL of generated droplets to the appropriate well on 96-well plate.
10. Once the addition of droplets to the plate is complete, seal the plate and amplify the droplets. Use the following parameters to set up the reaction: ramp rate 2 °C/s, activation at 90 °C for 10 min, then denaturation at 95 °C for 30 s followed by annealing at 55–65 °C for 1 min. Repeat denaturation and annealing steps for 40 cycles. Hold at 4–12 °C once complete. Note that the thermocycling parameters may need to be optimized based on experimental requirements.
11. Set up experiment in QuantaSoft Analysis Software provided by Bio-Rad. Designate well name, experiment type, and which assay corresponds to which fluorescent channels.
12. Run experiment and read droplets.
13. Analyze data using QuantaSoft Analysis Software. Refer to Droplet Digital PCR Applications Guide (http://www.bio-rad.com/webroot/web/pdf/lsr/literature/Bulletin_6407) for additional instructions on analysis methods (Fig. 1).

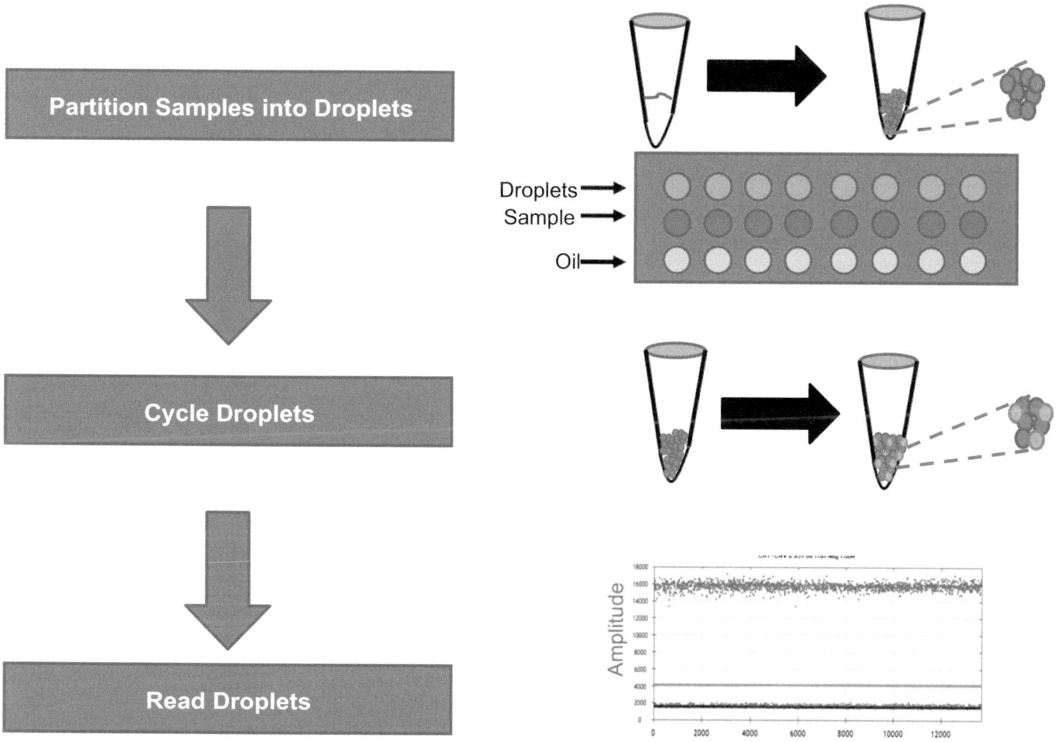

Fig. 1 Schematic flow through of ddPCR protocol. Samples are first partitioned into droplets. Droplets are then run through an end point PCR. Lastly, the samples are read using QX200 from Bio-Rad

5 Notes

1. Good-quality sample in the reaction ensures quality data. Therefore, samples should be of good quality and contamination free.
2. Primer designing should be carefully done to avoid self-annealing or cross-reactivity. Cross-reactivity might lead to off-target amplification.
3. As single nucleotide polymorphism (SNP) and rare mutation detection (RMD) detection requires careful evaluation of melting temperature (T_m) between the two probes, primer designing needs special attention.
4. All glassware and plasticware should be thoroughly cleaned to avoid contamination, as the technique is very sensitive.

References

1. Cao L, Cui X, Hu J et al (2017) Advances in digital polymerase chain reaction (dPCR) and its emerging biomedical applications. Biosens Bioelectron 90:459–474. https://doi.org/10.1016/jbios.2016.09.082

2. Quan PL, Sauzade M, Brouzes E (2018) dPCR: a technology review. Sensors (Basel) 18(4):E1271. https://doi.org/10.3390/s18041271
3. Yu M, Heinzerling TJ, Grady WM (2018) DNA methylation analysis using droplet digital PCR. Methods Mol Biol 1768:363–383. https://doi.org/10.1007/978-1-4939-7778-9_21
4. Taylor SC, Laperriere G, Germain H (2017) Droplet digital PCR versus qPCR for gene expression analysis with low abundant targets: from variable nonsense to publication quality data. Sci Rep 7(1):2409. https://doi.org/10.1038/s41598-017-02217-x
5. Harmala SK, Butcher R, Roberts CH (2017) Copy number variation analysis by droplet digital PCR. Methods Mol Biol 1654:135–149. https://doi.org/10.1007/978-1-4939-7231-9_9
6. Wang Y, Bergelson S, Feschenko M (2018) Determination of Lentiviral infectious titer by a novel droplet digital PCR method. Hum Gene Ther Methods 29(2):96–103. https://doi.org/10.1089/hgtb.2017.198
7. Mazaika E, Homsy J (2014) Digital droplet PCR: CNV analysis and other applications. Curr Protoc Hum Genet 82:7.24.1–7.2413. https://doi.org/10.1002/0471142905.hg0724s82
8. Gutierrez-Aguirre I, Racki N, Dreo T, Ravnikar M (2015) Droplet digital PCR for absolute quantification of pathogens. Methods Mol Biol 1302:331–347. https://doi.org/10.1007/978-1-4939-2620-6_24
9. Van Ginkel JH, Huibers MMH, Van Es RJJ, De Bree R, Willems SM (2017) Droplet digital PCR for detection and quantification of circulating tumor DNA in plasma of head and neck cancer patients. BMC Cancer 17(1):428. https://doi.org/10.1186/s12885-017-3424-0
10. Van Wesenbeeck L, Janssens L, Meeuws H, Lagatie O, Stuyver L (2018) Droplet digital PCR is an accurate method to assess methylation status on FFPE samples. Epigenetics 13(3):207–213. https://doi.org/10.1080/15592294.2018.1448679
11. Postel M, Roosen A, Laurent-Puig O, Taly V, Wang-Renault SF (2018) Droplet-based digital PCR and next generation sequencing for monitoring circulating tumor DNA: a cancer diagnostic perspective. Expert Rev Mol Diagn 18(1):7–17. https://doi.org/10.1080/14737159.2018.1400384
12. Perkins G, Lu H, Garlan F, Taly V (2017) Droplet-based digital PCR: application in cancer research. Adv Clin Chem 79:43–91. https://doi.org/10.1016/bs.acc.2016.10.001
13. Dobnik D, Stebih D, Blejec A, Morisset D, Zel J (2016) Multiplex quantification of four DNA targets in one reaction with bio-rad droplet digital PCR system for GMO detection. Sci Rep 6:35451. https://doi.org/10.1038/srep35451
14. Vossen RH, White SJ (2017) Quantitative DNA analysis using droplet digital PCR. Methods Mol Biol 1492:167–177. https://doi.org/10.1007/978-1-4939-6442-0_11
15. Parkin B (2019) Rare variant quantitation using droplet digital PCR. Methods Mol Biol 1881:239–251. https://doi.org/10.1007/978-1-4939-8876-1_18
16. Otsuji K, Sasaki T, Tanaka A et al (2017) Use of droplet digital PCR for quantitative and automatic analysis of the HER2 status in breast cancer patients. Breast Cancer Res Treat 162(1):11–18. https://doi.org/10.1007/s10549-016-4092-5

Chapter 14

Determination of Michaelis–Menten Enzyme Kinetics Parameters of Alkaline Phosphatase in Clinical Samples

Varsha Gupta, Abhishek Gupta, Lata I. Shukla, Abhimanyu Kumar Jha, Jaya Prakash, and Baishnab Charan Tripathy

Abstract

Enzymes are specific biocatalysts which enhance the rate of biological reactions in mild conditions of temperature, pressure and a particular pH and thus have an important role in almost every cellular process. They play an indispensable role in metabolism, signal transduction cascade, and gene expression which makes them an important target for drug development, bioprocess industry, food industry, and eco-friendly green chemistries. They are emerging as potential drug targets. Many diseases are been explored where enzyme inhibition can have therapeutic effects. The study and analysis of enzyme have been challenging due to its mathematical aspects. However, due to their important functions, it becomes imperative to have sufficient and thorough knowledge of biochemistry of an enzyme and deep understanding of kinetics of enzyme action. In the present chapter, we will explore very simple and established basic concept of enzyme kinetics, estimation of initial reaction velocities with different substrate concentration, measurement and utility of K_M and V_{max}, catalytic power of the enzyme along with its turnover number. The readers will learn about changes which occur in their graph with different enzyme inhibitors in their reaction. This chapter would also be helpful for researchers and students to understand positive and negative feedback loop. We have provided example of serum alkaline phosphatase (ALP) where one can learn to establish different parameters with increase in absorbance as substrate concentration increases.

Keywords Enzyme kinetics, Michaelis–Menten equation, Lineweaver–Burk plot, Eadie–Hofstee plot, Alkaline phosphatase, Allosteric regulation

1 Introduction

Enzymes are biocatalysts which lower activation energy or Gibbs free energy (ΔG) of the system (Fig. 1). The biological molecules or substrates have requirement of high activation energy for conversion into products. This is essentially important to avoid spontaneous formation of products by different reactants. Thus, enzymes are important to bring about the reaction and have better control over metabolic processes. Enzymes lower activation energy so that reactants can come together, overcoming forces of repulsion which induce formation and/or breaking of bonds to form the product.

Neelima Gupta and Varsha Gupta (eds.), *Experimental Protocols in Biotechnology*, Springer Protocols Handbooks, https://doi.org/10.1007/978-1-0716-0607-0_14,

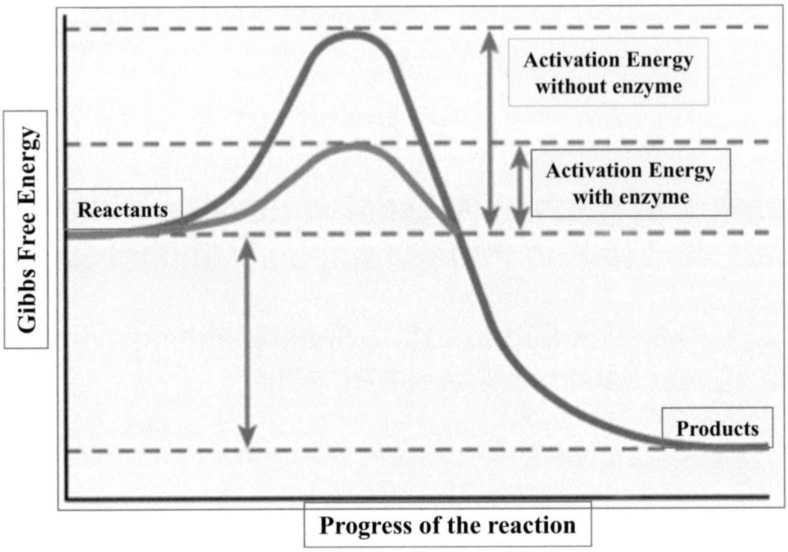

Fig. 1 The Gibbs free energy changes with and without enzymes. The blue curve shows requirement of high energy for reaction, while green curve shows the lowering of activation energy using enzyme

Measurement of rates of catalytic reactions is important for establishing enzyme activity. It helps to determine rate changes of enzymatic reactions in different conditions. The rate changes also help to establish the role of different inhibitors or activators on enzyme action, which is finding increasing importance in drug discovery. Now enzymes are produced, modified, optimized, and explored for their usefulness in several fields. In this chapter, readers would learn about Michaelis and Menten kinetics of enzyme, concept of enzyme inhibition, which is finding increasing importance in therapy.

2 Technique

The chapter explores the enzymatic assay of serum alkaline phosphatase (ALP) where different kinetic parameters can be established by using absorbance value and extinction coefficient with known substrate concentration for calculation of K_M and V_{max}. The ALP is an enzyme which is found in many tissues of the human body. It is in ample amount in liver and bone cells. It also plays an important role in growth and development of the teeth and bones. The normal range for alkaline phosphatase is 44–147 IU per liter, in humans. Higher serum levels indicate liver disease, bone disorders, leukemia, and lymphoma. Low levels of alkaline phosphatase may be associated with malnutrition, Wilson's disease, and protein deficiency. The ALP is used to monitor alcoholism, renal cell carcinoma, pancreatitis, and gallstones. The catalytic site of ALP consists of a two-metal-ion (bimetallo) Zn^{2+} catalytic core which stabilizes the transition state.

2.1 Enzyme Kinetics

Consider a reaction S → P, in which the substrate (S) converts into the product (P). In a biological reaction the change of substrate into the product can either occur in a single step or may involve several intermediates. The reaction is facilitated by biological catalysts called enzyme. The substrate binds to the enzyme to form the enzyme–substrate complex which then converts into the product. Enzymes are not used up in the reaction [1].

$$\mathrm{E} + \mathrm{S} \underset{k_{-1}}{\overset{k_1}{\rightleftharpoons}} \mathrm{ES} \underset{k_{-2}}{\overset{k_2}{\rightleftharpoons}} \mathrm{P} + \mathrm{E}$$

The rate at which substrate disappears or product is formed is called velocity (Vo) of the reaction. Rate of change of S to P would be sum of rate constants k which is dependent upon enzyme involved in the reaction, and concentration of starting material [S] where, k_1 is the rate constant for the formation of [ES] complex and k_{-1} is the rate constant of dissociation of [ES] complex to form free enzyme [E] and substrate [S]; and k_2 is the rate constant for the conversion of the [ES] complex into product [P] and enzyme [E], while k_{-2} would lead to formation of [ES] complex.

The condition is known as steady state of the reaction where rate of formation of ES is equal to the rate of breakdown of ES. This occurs at high substrate concentration, when [ES] levels are constant.

Cofactors: *Many biological enzymes require the presence of small molecules for catalyzing the reactions. These small molecules can be referred to as cofactors. The enzyme without cofactor is termed as apozyme, while the complete enzyme with its cofactor is called holozyme.*

$$\text{Apozyme} + \text{Cofactor} \rightarrow \text{Holozyme}$$

Cofactors can be metal ions or small organic compounds. Metals as Zn^{2+} (carboxypeptidase), Mg^{2+} (many restriction enzymes), and Ni^{2+} (urease) can act as a cofactor. Major small organic molecules which act as cofactor are derived from vitamins. These vitamins derived small organic molecules such as thiamine pyrophosphate (pyruvate dehydrogenase) and nicotinamide adenine dinucleotide (lactate dehydrogenase) are referred to as coenzymes. These coenzymes may be either tightly bound (known as prosthetic group) or *loosely bound. Loosely bound coenzymes are more likely to bind the enzyme as a cosubstrate* [1].

Reaction order: *The velocity of reaction is directly proportional to the concentration of the substrate. This would be first order reaction. Thus, the order of reaction essentially corresponds to the molecularity, that is, the number of starting molecules which must come together or collide to generate the product. If two molecules are involved in the reaction, it will be second order reaction* [2].

Rate1 = k_1 [E] [S] (concentration of starting material E and S)
Rate2 = k_2 [ES] (concentration of starting material ES)
Thus, rate would be velocity of the reaction (V_0)

$$\text{Rate} = V_0 = \frac{[\mathrm{d}P]}{\mathrm{d}T} = \frac{\Delta P}{\Delta T}$$

Therefore, velocity of the reaction would be rate of change of concentration of the product with respect to time. **Rate of the reaction can be increased by**

1. Increasing concentration of substrate.

   or

2. Increasing concentration of enzyme.

However, in biological reaction, concentration of enzyme is constant, now if enzyme is free and substrate concentration is low, the reaction rate can be increased by increasing the concentration of substrate. According to this approach, the rate (velocity) of enzyme-catalyzed reaction is measured with different substrate concentrations and the initial rate of reaction is dependent on substrate concentration [S]. It means the initial velocity (V_0) increases linearly as the substrate concentration is increased. Gradually the velocity increases with increase in substrate concentration and then reaches maximum velocity (V_{max}); after this it does not further increase when the substrate concentration increases. As with increasing substrate concentration, all active sites of the enzyme are blocked, there will be no further increase in rate of reaction as there is no free enzyme left. Then the reaction would have maximum rate, that is, V_{max} (that is, all available enzymes are saturated with substrate molecules).

Maximum velocity: *The enzyme concentration is fixed in a particular reaction; thus, initially when active sites of the enzyme are free, the rate of the reaction can be enhanced by increasing the concentration of the substrate. But when substrate concentration becomes high, such that, all active sites of enzyme are occupied.*

There will be no further increase in the velocity of the reaction at increasing substrate concentration as all the active sites of the enzymes are occupied. At this stage, the maximum velocity is achieved which is not affected by further increase in substrate concentration.

As solutions are behaving ideally thus, in steady state, rate of forward and backward reaction with respect to intermediate [ES] should be balanced (in this, conversion of S to P without enzyme is negligible thus, can be ignored).

Steady state assumes that [ES] is constant at high substrate concentration.

Formation of ES = Loss of ES

$$E + S \underset{-1}{\overset{1}{\rightleftharpoons}} ES \underset{-2}{\overset{2}{\rightleftharpoons}} P + E$$

Therefore, Rate 1 + Rate − 2 = Rate − 1 + Rate 2

In this, one should remember that products normally do not convert back into reactants as all enzymatic reactions are thermodynamically stable. Thus, we can ignore rate −2. Putting it as rate forming ES = rate of loss of ES

$$E + S \underset{k_{-1}}{\overset{k_1}{\rightleftharpoons}} ES \overset{k_2}{\rightleftharpoons} P + E$$

$$k_1[E][S] = k_{-1}[ES] + k_2[ES] \tag{1}$$

However, we cannot measure [ES] and [E] but we can measure total enzyme concentration $[E]_T$. Total enzyme available $[E]_T = [E] + [ES]$ (total enzyme available would be equal to free enzyme + enzyme bound with substrate.

Substituting this in Eq. (1) $([E]_T - [ES]) = [E]$

$$k_1([E]_T - [ES])[S] = [ES](k_{-1} + k_2)$$

dividing the equation by k_1 yields

$$[E]_T[S] - [ES][S] = [ES]\left(\frac{k_{-1} + k_2}{k_1}\right) \tag{2}$$

$\left(\frac{k_{-1}+k_2}{k_1}\right) = K_M$ (K_M is Michaelis constant)

Rearranging and substituting values in Eq. (2) gives

$$\begin{aligned} [E]_T[S] &= K_M[ES] + [ES][S] \\ = [E]_T[S] &= [ES]\,(K_M + [S]) \end{aligned} \tag{3}$$

Divide the Eq. (3) with $K_M + [S]$

$$[E]_T[S]/K_M + [S] = [ES] \tag{4}$$

$V_0 = [dP]/dT = \Delta P/\Delta T$ (rate of reaction is equal to product formation, i.e., k_2[ES])

$$V_0 = k_2[\mathrm{ES}]$$

here both $[\mathrm{E}]_\mathrm{T}$ and [S] can be measured, however, to meet the condition of steady state, the substrate concentration should be high so that each enzyme molecule is bound to substrate and converts it into product keeping [ES] constant. Thus, velocity is equal to rate of formation of our product, that is, rate 2 or k_2[ES]. Now using this equation, we multiply both sides of Eq. (4) with k_2.

Therefore,

$$k_2[\mathrm{ES}] = k_2[\mathrm{E}]_\mathrm{T}[\mathrm{S}]/K_\mathrm{M} + [\mathrm{S}] \quad (5)$$

If the reaction has approached maximum velocity, then $V_0 = V_{\max}$, that is the available enzyme $[\mathrm{E}]_\mathrm{T}$ is bound to substrate [ES] and there will be no free enzyme left. Therefore, k_2[ES] would be equal to $k_2[\mathrm{E}]_\mathrm{T}$, as the reaction is at its maximum rate,

Thus, $k_2[\mathrm{E}]_\mathrm{T} = V_{\max}$ instead of being equal to V_0.

Therefore, substituting these values in Eq. (5) will yield Michaelis–Menten equation; (If $V_0 = V_{\max}$ then $[\mathrm{E}]_\mathrm{T} = [\mathrm{ES}]$ maximum velocity would only be possible if at high substrate concentration all the enzyme is saturated with substrate, thus, k_2 $[\mathrm{E}]_\mathrm{T} = V_{\max}$).

$V_0 = \frac{V_{\max}[\mathrm{S}]}{K_\mathrm{M}+[\mathrm{S}]}$ Michaelis–Menten equation; K_M—Michaelis constant [3].

If we plot a graph between initial velocity (V_0) versus substrate concentration [S], this is called Michaelis–Menten curve. The curve describes the relationship between constant enzyme concentration and the variable substrate concentration. Michaelis and Menten gave a mathematical equation which established the relationship between initial velocity (V_0), substrate concentration [S], maximum velocity ($V_{\max}$), and Michaelis constant (K_M).

$$V_0 = \frac{V_{\max}[\mathrm{S}]}{K_\mathrm{M} + [\mathrm{S}]}$$

Now, assume K_M is the substrate concentration; substituting $K_\mathrm{M} = [\mathrm{S}]$

$$V_0 = \frac{V_{\max}[\mathrm{S}]}{[\mathrm{S}] + [\mathrm{S}]} = \frac{V_{\max}[\mathrm{S}]}{2[\mathrm{S}]} = \frac{V_{\max}}{2}$$

When K_M is equal to the substrate concentration, the reaction velocity becomes half of its maximum velocity (Fig. 2).

For the determination of K_M and $V_{\max}$, we need to determine a set of V values at various substrate concentrations [S] [4], and plot the graph (Fig. 2). The Michaelis–Menten equation describes how the reaction velocity varies with substrate concentration. K_M is numerically equal to the substrate concentration at which the

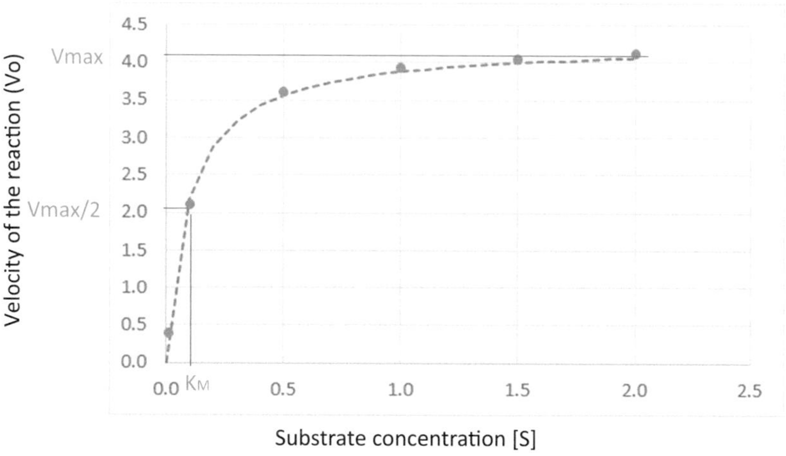

Fig. 2 Plot of Michaelis–Menten equation. The plot is made between substrate concentration on *x*-axis and velocity (V_0) on *y*-axis. The curve where saturation occurs at high substrate concentration is the V_{max} and $V_{max}/2$ gives K_M values

reaction velocity is equal to $1/2\,V_{max}$, where V_{max} is maximum velocity at which reaction is independent of [S] concentration. The lower the K_M, the better the working of enzyme when substrate concentrations are small. K_M can be used to analyze the ability of the enzyme to catalyze the reaction. Thus, concentration of substrate at which the velocity becomes half of the maximum velocity is called Michaelis constant (K_M).

At low concentration of substrate ($[S] \ll K_M$), the velocity of the reaction is first order and it is proportional to substrate concentration ($V_0 \propto [S]$) while at high concentration of substrate ($[S] \gg K_M$), the velocity of the reaction is zero order as it is independent of substrate concentration, that is, $V_0 = V_{max}$ [5].

2.2 Determining Catalytic Efficiency and Turn Over Number

In Fig. 2, initially the graph shows linear increase, indicating that the velocity is directly proportional to substrate concentration. As substrate concentration is further increased, velocity does not increase linearly with increasing substrate concentration. At a particular stage when there is no further increase in velocity, that is the maximum possible rate at which the enzyme (at that particular concentration) can operate on that particular substrate [1].

V_{max} = product of turnover number $K_{cat} \times [E]_T$

All the active sites in an enzyme are saturated with substrate.

K_M can be used to define catalytic power of an enzyme which is referred to as catalytic efficiency. It is essential to measure enzymatic efficiency.

$$K_{cat} = V_{max}/[E]_T$$

Catalytic efficiency $= K_{cat}/K_M$; high K_{cat} and lower K_M indicate increase in catalytic efficiency of an enzyme.

In the physiological condition, substrate concentration is not very high but is low, normally between 0 and K_M value.

$$0 < [S] < K_M$$

Thus, concentration of substrate is much smaller than the K_M value.

Therefore, physiological reactions occur at much lower velocity than maximum reaction velocity. In enzyme kinetics, it is important to determine that how many substrate molecules an enzyme can convert into product per catalytic site per unit time.

$$K_{cat} = V_{max}/[E]_T$$

Turnover number: It shows how many substrate molecules are converted into product in 1 s, that will be maximum speed of the reaction, $K_{cat} = s^{-1}$, (Turnover number (kcat) = (moles of product/sec)/ (moles of enzyme) or sec^{-1})

2.3 Alkaline Phosphatase

Alkaline phosphatase (ALP, EC 3.1.3.1) activity test measures the amount of ALP in the serum. ALP is ubiquitous enzyme found in almost every organ of the body as liver, bones, kidneys, and digestive system. In humans, ALP consists of five tissue specific enzymes. These tissue specific enzymes can be resolved by gel electrophoresis. Serum ALP, ideally represent severity of the disease in organ or tissue of their origin, thus, it has gained clinical interest. These include bone disorders, rickets, obstructive jaundice which were shown to have high levels of ALP in the plasma. ALP levels are high in tissues active in transport of nutrients, suggesting its involvement in nutrients transport across the epithelial membrane. High serum ALP normally predicts liver or bone diseases.

There are several other types of blood tests that are used to evaluate liver function. These include bilirubin, aspartate aminotransferase (AST), and alanine aminotransferase (ALT) tests. If these results are normal and the alkaline phosphatase levels are high, it may mean the problem is not in the liver. Instead, it can indicate a bone disorder, such as Paget's Disease of Bone, a condition that causes the bones to become abnormally large, weak, and prone to fractures. Moderately high levels of alkaline phosphatase may indicate conditions such as Hodgkin lymphoma, heart failure, or a bacterial infection. Low levels of alkaline phosphatase may indicate hypophosphatasia, a rare genetic disease that affects bones and teeth. Low levels may also be due to a deficiency of zinc or malnutrition [4, 6, 7].

3 Materials

Material: The following material would be required for conducting assay of Alkaline phosphatase from serum samples

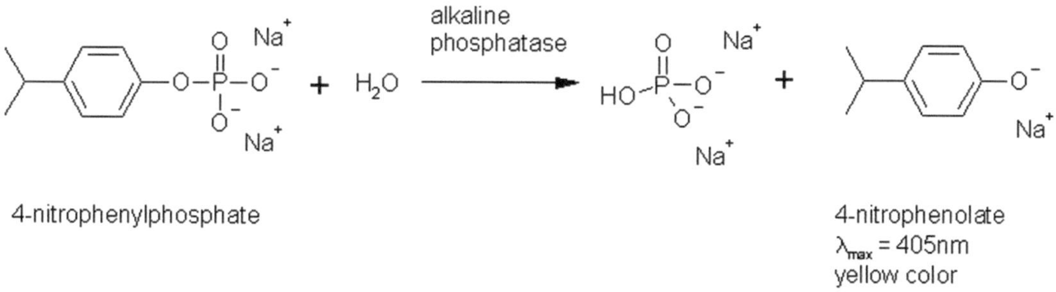

Fig. 3 Alkaline phosphatase catalyze the conversion of colorless 4-nitrophenyl phosphate to 4-nitrophenoxide ion that shows intense yellow color with molar absorptivity at 405 nm. Thus, levels of 4-nitrophenoxide ions are detected using spectrophotometer with monitoring with respect to time for enzyme assay

- Serum samples, pipette tips, reagent bottles, distilled water, tissue rolls, alkaline phosphatase enzyme, 4-nitrophenyl phosphate, diethanol amine buffer (DEA), $MgCl_2$, KCl, HCl

4 Methods

Principle: ALP can hydrolyze many organic monophosphate esters leading to the formation of alcohol or phenol and free phosphate (Pi) ion. Subsequently, the phosphate group may be transferred to water or other acceptors of phosphate present. Under optimal conditions, ALP can catalyze two transphosphorylation reactions (Fig. 3).

$$\text{4-Nitrophenyl phosphate} + H_2O \xrightarrow{\text{Alkaline phosphatase}} \text{4-Nitrophenoxide} + \text{Pi}$$

(H_2O serves as phosphate acceptor)

4-nitrophenyl phosphate is colorless at experimental pH of 10.4 while 4-nitrophenoxide ion exhibits intense yellow color with molar absorptivity at 405 nm. Thus, levels of 4-nitrophenoxide ions are detected using spectrophotometer where monitoring with respect to time is done for enzyme assay [4, 6, 7].

4.1 ALP Assay and Michaelis–Menten Plot

The assay can be conducted using DEA buffer or Tris–HCl buffer. Prepare 1.0 M DEA buffer and add 0.50 mM $MgCl_2$, adjust pH 9.8 with 5 M HCl. Assay can also be conducted using alkaline phosphatase buffer containing 100 mM Tris pH 9.5, 100 mM NaCl, 5 mM $MgCl_2$. The substrate used is 4-nitrophenyl phosphate as it is readily hydrolysable. Upon hydrolysis it yields chromogenic product with molar absorptivity of 18.5 $mMol^{-1}$ cm^{-1} and, gives high sensitivity with small sample volumes. Substrate 4-nitrophenyl phosphate can be used with 30 mM working solution, prepare fresh and use within 8–12 h, store in dark bottle. Prepare Alkaline phosphatase enzyme with 0.15 units/mL, reconstitute with assay buffer.

4.2 Protein Estimation of Samples

For protein estimation and equalization, please refer to Chapter 8 (Subheading 4.3).

4.3 Standard Curve

Prepare the standard curve as described: Switch on UV-visible spectrophotometer 30 min prior to start of reaction. Use quantitative mode with absorbance measurement at 405 nm. After protein estimation of samples (use fix amount of protein from all the experimental samples), make all solutions and stocks ready. Take both the cuvettes and add buffer and enzyme in the both cuvettes for preparing substrate blank [or prepare enzyme blank or buffer blank (different blanks help to establish authenticity of data)] and for blanking the instrument. These cuvettes would have all the reagents except substrate. Put them in spectrophotometer and do blanking or Auto zero.

Now stepwise add buffer and substrate in the sample cuvette (do not disturb the blank cuvette), add enzyme, mix well and start recording the absorbance till 3 min with 15/30 s interval. Measure gradually for all the standards.

The reaction is for 3 mL (Table 1)

Initial and end absorbance for each standard with respect to time is obtained, time observed for linear increase is recorded (take that time only). Plot absorbance versus time to obtain a curve as represented in Fig. 4. This would be absorbance versus time curve (Fig. 4).

Now calculate slope and velocity by using the formula:

$$= \text{change of absorbance/change in time} = \Delta A/\Delta t$$

Example: Absorbance change for first standard is 0.26 and linear time change 0.75 s. Using $\Delta A/\Delta t = 0.26/0.75 = 0.35\ \text{min}^{-1}$. For each standard value, calculate the rate of change of absorbance.

Now it is more appropriate to use the substrate concentration, using Beer–Lambert law

$$A = c \times \varepsilon \times l$$

Change in absorbance (A) = concentration (c) × extinction coefficient (ε) × length (l) (Table 2).

Extinction coefficient (ε) for alkaline phosphatase $=18.5\ \text{mM}^{-1}\ \text{cm}^{-1}$; length of cuvette is 1 cm [4].

The above is for one standard, accordingly all standards as shown in Table 3 can be calculated. Concentration of 18.9 mM, that is, mM per litre, convert it to nM/mL. Total reaction volume

Table 1
The components used for plotting standard curve

Concentration	Buffer (μL)	Substrate 4-NPP (μL)	Standard enzyme (μL)
Blank	2950	0	50
Standard1	2940	10	50
Standard2	2930	20	50
Standard3	2900	50	50
Standard4	2850	100	50
Standard5	2800	150	50
Standard6	2750	200	50
Standard7	2700	250	50
Standard8	2650	300	50
Standard9	2600	350	50
Standard10	2550	400	50

The total volume is 3 mL with reaction consisting of buffer, substrate and enzyme. Enzyme is added just before recording of absorbance

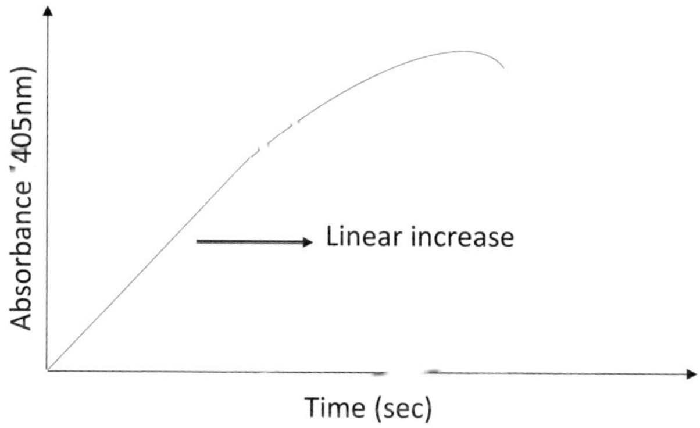

Fig. 4 The plot of absorbance versus time of a standard. With increasing amount of substrate, there would be increase in absorbance with respect to time for alkaline phosphatase assay

was 3 mL; therefore, calculate total product formed in 3 mL reaction (Tables 2 and 3).

Total product is plotted on *y*-axis and final substrate concentration on *x*-axis, a rectangular hyperbolic curve (Fig. 5) is obtained. When the stationary phase is observed, that is, the reaction velocity in not further increasing with increase in substrate concentration that is, V_{max} (maximum reaction velocity). K_M can be calculated by the curve as shown in Fig. 5. The Michaelis constant is the substrate concentration at which reaction velocity (V) is half of its maximum

Table 2
Shows the formation of actual product using absorbance

Slope V $\Delta A/\Delta t$	Calculating the substrate concentration by Beer–Lambert law	Converting the units to μMoles/min (×1000)	Convert mM/l to nM/mL	Calculate total product formed in 3 mL (*y*-axis)	Calculating final substrate concentration (*x*-axis)
$\Delta A/\Delta t$	$C = A/\varepsilon \times l$ Substituting $A/$ 18.5 mM^{-1} cm^{-1} 1 cm	Micromoles/ min	nM/mL	For 3 mL reaction volume (×3)	We have used 10 μl of 30 mM working stock in 3 mL reaction
0.26/0.75 = 0.35 min^{-1}	=0.35 min^{-1}/ 18.5 mM^{-1} cm^{-1} =0.0189 mM/min	18.9 μM/min	18.9 nM/min	56.7 nmoles/min	Final concentration = 0.1 mM

Table 3
The final product concentration formed at different concentrations of substrate

Concentration (mM)	Velocity (nM/min)
0	0
0.1	56.7
0.2	94
0.5	153
1	201
1.5	265
2	316
2.5	368
3	435
3.5	441
4	441

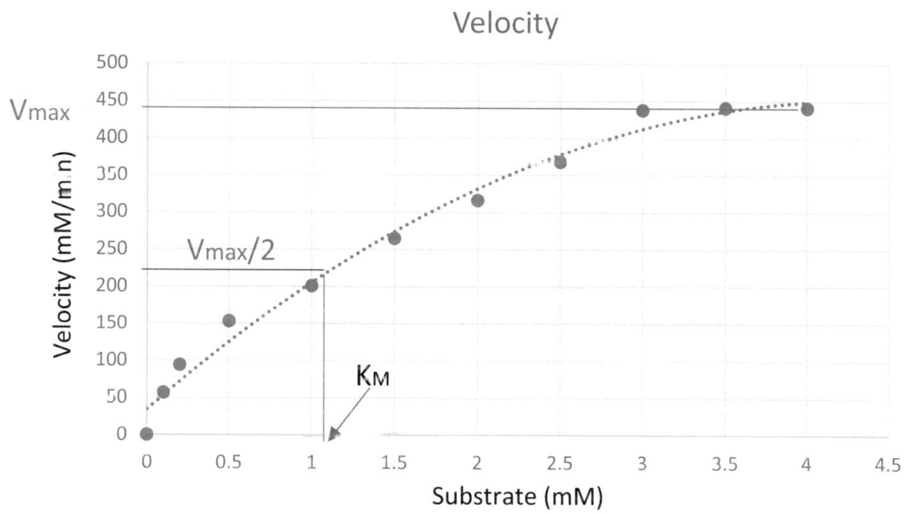

Fig. 5 Substrate concentration on *x*-axis and product formed on *y*-axis (plotted from data of Table 3). In this graph, when there is no further increase in product formation with increasing substrate concentration, that would be V_{max}. The Michaelis constant is the substrate concentration at which reaction velocity (V_0) is half of its maximum velocity ($V_{max}/2$)

velocity ($V_{max}/2$). Thus, if values for K_M are small, the enzyme shows high catalytic efficiency. K_M is unique for each enzyme–substrate pair and is also dependent on pH and temperature.

It is important to express values in terms of activity of the enzyme. Here, enzyme activity is calculated as moles of product formed per minutes. Enzyme unit is defined as the amount of enzyme that can catalyze the reaction of 1 μmole of substrate per minute (standard unit) or I nmole of substrate/minute is 1 unit (IU).

The activity of the enzyme can be calculated as μmoles of product formed per minute per mL or nmoles of product formed/minute/mL.

K_{cat}/K_M is a measure for catalytic efficiency.

$$K_{cat} = V_{max}/[E]_T$$

4.3.1 Specific Activity of Enzyme

Biological fluids are complex which have mixtures of different proteins. It is important to determine the presence of particular protein or particular enzyme in the biological fluids. Specific activity helps us determine that how much enzyme is present in our body fluids. Thus, the degree of purity of an enzyme is represented by the specific activity of the enzyme.

Specific activity can be defined as the amount of product that is formed by the enzyme (in a particular time) per mg/mL of total protein. Therefore, for calculation of specific activity, determination of enzyme activity (*see* Subheading 4.3) and protein concentration (*see* Chapter 8, Subheading 4.3) is required.

Activity of the enzyme can also be presented as enzyme unit, that is, concentration of product formed (nmoles) per minute per mL (is defined as 1 unit). It may be represented as micromoles per minute (μmol/min) or nanomoles per minute (nmol/min) (rate of reaction)

Thus, it is the ratio of enzyme activity divided by concentration of protein.

Specific activity is the amount of product produced by the activity of an enzyme in the given time (nmoles/min/mL) per milligram of total protein of the sample. Amount of product formed (enzyme unit) can be given as μmoles/min or nmoles/min (*see* Subheading 4.3).

Total protein can be expressed as mg/mL

$$\text{Specific activity} = \frac{\text{Enzyme activity}}{\text{Protein concentration}}$$
$$= \text{units/mL divided by protein concentration in mg/mL} = \text{nmoles/min/mg}$$

4.4 Alternative Version of Michaelis–Menten Plot

4.4.1 Lineweaver Burk Plot

We have discussed about Michaelis–Menten kinetics, there the exact determination of V_{max} from the curve is difficult (*see* Fig. 5). Different scientists have tried to rearrange the Michaelis–Menten

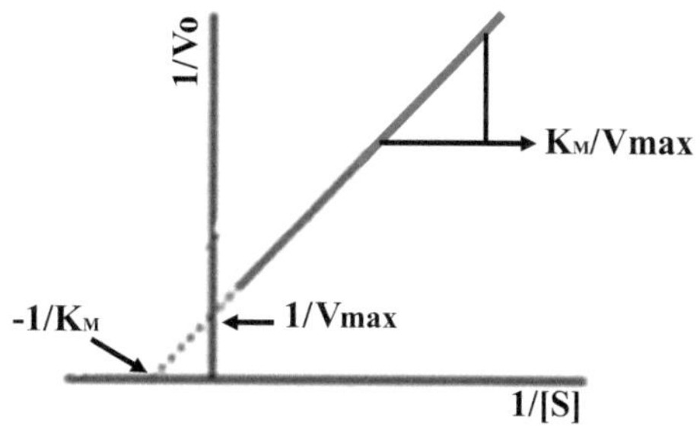

Fig. 6 Double reciprocal plot between 1/[S] and $1/V_0$. *Y*-intercept is $1/V_{max}$, $-1/K_M$ at *x*-intercept and slope is K_M/V_{max}

equation for easy and appropriate determination of V_{max}. In Lineweaver–Burk plot or double reciprocal plot, determination of V_{max} and K_M is accurate as we obtain a straight line [8].

In Lineweaver–Burk plot we take the reciprocal of the Michaelis–Menten equation.

$V_0 = \frac{V_{max}\,[S]}{K_M+[S]}$ (Michaelis–Menten equation)

Now taking reciprocal of this equation yields

$$\frac{1}{Vo} = \frac{K_M + [S]}{V_{max}\,[S]}$$

$\frac{1}{Vo} = \frac{K_M}{V_{max}}\frac{1}{[S]} + \frac{1}{V_{max}}$. This equation is similar to the equation of the straight line

$$y = mx + b$$

$y = mx + b$ (if we plot a graph between y and x, the slope would be m and the intercept would be b).

Thus, plotting graph between $\frac{1}{Vo}$ and $\frac{1}{[S]}$, the slope would be $\frac{K_M}{V_{max}}$ and the y-intercept would be $\frac{1}{V_{max}}$ and the x-intercept would be $-\frac{1}{K_M}$.

In this graph we get a linear line (Fig. 6) in place of a curved graph that is obtained with Michaelis–Menten equation (Fig. 5).

If we extrapolate the line further it cuts the x-axis; when it cuts the x-axis (Fig. 5) the value of y is zero, substituting the value of $y = 0$ in the double reciprocal equation

$0 = \frac{K_M}{V_{max}}\frac{1}{[S]} + \frac{1}{V_{max}}$ rearrange the equation

$-\frac{1}{V_{max}} = \frac{K_M}{V_{max}}\frac{1}{[S]}$ rearrange further

$-\frac{1}{K_M} = \frac{1}{[S]}$ Thus, when $y = 0$ then the point where the line intersects the x-axis is $= -\frac{1}{K_M}$.

Lineweaver Burk Plot and Enzyme Inhibition

Lineweaver–Burk plot or double reciprocal plot is very useful when you wish to determine the effects of different inhibitors like competitive, uncompetitive, and noncompetitive on enzyme activity.

Competitive Inhibitors

The competitive inhibitor binds at the site where the substrate binds; thus, it competes with the substrate molecule to bind at an active site. If a competitive inhibitor is added to the reaction, and the double reciprocal plot is made. If after the addition of an inhibitor, extrapolated point which intersects at *x* axis shifts to the right, that is, K_M is increased (as it is reciprocal) and V_{max} remains same, it indicates the presence of an inhibitor which does not influence the maximum velocity of the reaction. However, it increases the substrate concentration required to achieve ½ of the maximum velocity. This would be a competitive inhibitor.

Therefore, this inhibitor would increase the K_M value (a shift toward right as it is inverse value), but the V_{max} value will remain the same. A high substrate concentration would be required for obtaining $V_{max}/2$. The slope also increases in the presence of competitive inhibitor (Fig. 7a)

Uncompetitive Inhibitors

This inhibitor can bind with an enzyme–substrate complex and thus does not compete with the substrate for the binding site. Double reciprocal plot after addition of an uncompetitive inhibitor: If after the addition of an inhibitor, the extrapolated point which intersects at *x* axis shifts to the left, that is, K_M is decreased (as it is reciprocal) and V_{max} is decreased, it indicates the presence of the inhibitor which influences K_M and the maximum velocity of the reaction. This would be an uncompetitive inhibitor. This results in a decrease of both K_M and V_{max} values. The presence of an uncompetitive inhibitor would result in the appearance of two parallel lines in the double reciprocal plot (Fig. 7b).

Noncompetitive Inhibitor

Noncompetitive inhibitors can bind the enzyme before the binding of the substrate. Thus, they will cause a decrease in V_{max} value, while the K_M is unaffected. Therefore, addition of this inhibitor would result in changes in the position on the graph of the line where it intercepts *y*-axis, that is, V_{max} would decrease and *x*-intercept remains the same for the initial enzyme reaction before and after the addition of the inhibitor, and K_M remains unchanged, while *y*-intercept increases, showing decrease in V_{max} with a large slope (Fig. 7c).

The inhibitor changes $1/K_M$ but changes do not occur in V_{max}—Competitive inhibitor

The inhibitor changes V_{max} but does not influence K_M—Noncompetitive inhibitor

The inhibitor changes both K_M and V_{max}—Uncompetitive inhibitor

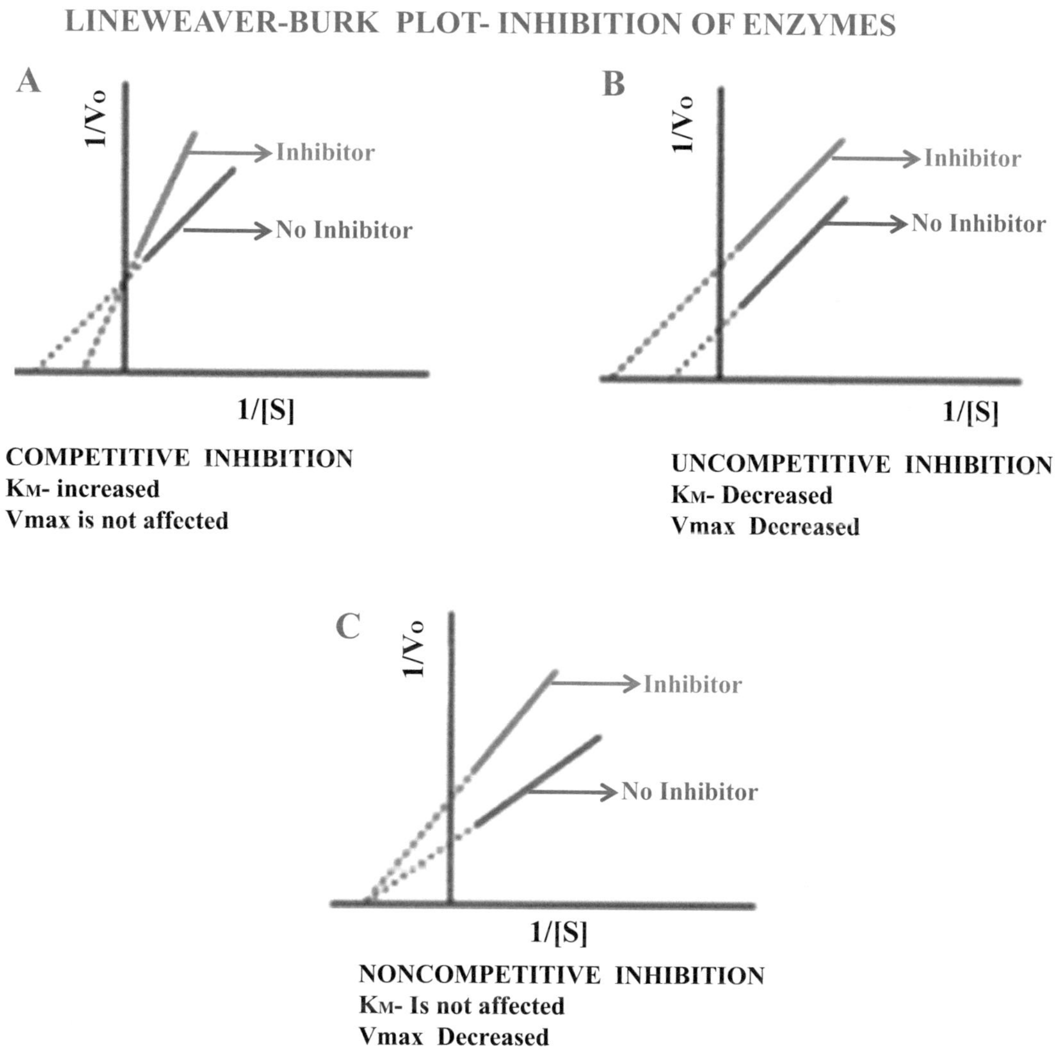

Fig. 7 (**a**) If competitive inhibitor is added to the reaction and double reciprocal plot is made, the inhibitor increase the K_M value (shift towards right as it is inverse value) but V_{max} value will remain same. (**b**) In case of uncompetitive inhibitor, intercept at x axis shifts to the left, i.e., K_M is decreased (as it is reciprocal) and V_{max} is also increased, it indicates presence of inhibitor which is influencing K_M and maximum velocity of the reaction. (**c**) Noncompetitive inhibitor will cause a decrease in V_{max} value, while the K_M will be unaffected. Therefore, addition of this inhibitor would result in changes in the position on the graph of the line where it intercepts y-axis, i.e., V_{max} would decrease

4.4.2 Eadie–Hofstee Plot

The Eadie–Hofstee plot is a precise linear plotting method by further rearrangement of Michaelis–Menten equation. Eadie–Hofstee plot also uses rearranged Michaelis–Menten equation where V_0 is plotted against $V_0/[S]$ [9, 10].

$$V_0 = \frac{V_{max}\,[S]}{K_M + [S]} \text{ (Michaelis–Menten equation)}$$

Rearranging this equation

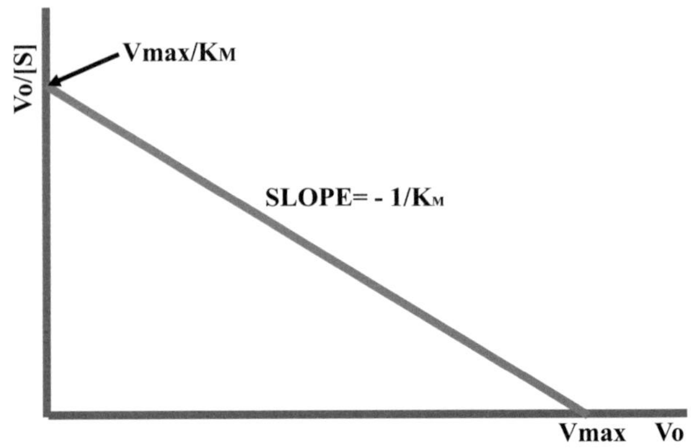

Fig. 8 The Eadie–Hofstee plot between V_0/[S] at *y*-axis and V_0 at *x*-axis is shown. The point at which it cuts *x*-axis is V_{max}, slope is $-1/K_M$, and point at which it cuts *y*-axis is V_{max}/K_M

$$V_0\,(K_M + [S]) = V_{max}[S] = V_0\frac{(K_M + [S])}{[S]} = V_{max}$$
$$= V_0\frac{K_M}{[S]} + V_0\frac{[S]}{[S]} = V_{max} = \frac{V_0}{[S]}\,K_M + V_0$$
$$= V_{max} = \frac{V_0}{[S]}K_M = V_{max} - V_0 = \frac{V_0}{[S]}$$
$$= \frac{V_{max} - V_0}{K_M} = \frac{V_0}{[S]} = \frac{V_{max}}{K_M} - \frac{V_0}{K_M} = \frac{V_0}{[S]}$$
$$= V_0\,(-\frac{1}{K_M}) + \frac{V_{max}}{K_M}, y$$
$$= xm + b \text{ equation of straightline}$$

The final equation looks like straight line equation, y=xm+b. The graph is plotted between y ($V_0/[S]$) and x (V_0) is shown in Fig. 8. The point at which it cuts *x*-axis is V_{max}, slope is $-1/K_M$ and point at which it cuts *y*-axis is V_{max}/K_M.

Therefore, point when it cuts *x*-axis, (*see* Fig. 8), value for *y*-axis is zero, putting this value in the above equation.

$$= 0 = V_0\left(-\frac{1}{K_M}\right) + \frac{V_{max}}{K_M}$$
$$= \frac{V_0}{K_M} = \frac{V_{max}}{K_M}$$

$= V_0 = V_{max}$ (at zero value of *y*), here V_{max} determination is easier than double reciprocal plot and Michaelis–Menten plot.

The Eadie–Hofstee plot is useful to determine important kinetic parameters but both the ordinate and abscissa in this plot are dependent variables which might lead to experimental errors on both the axis.

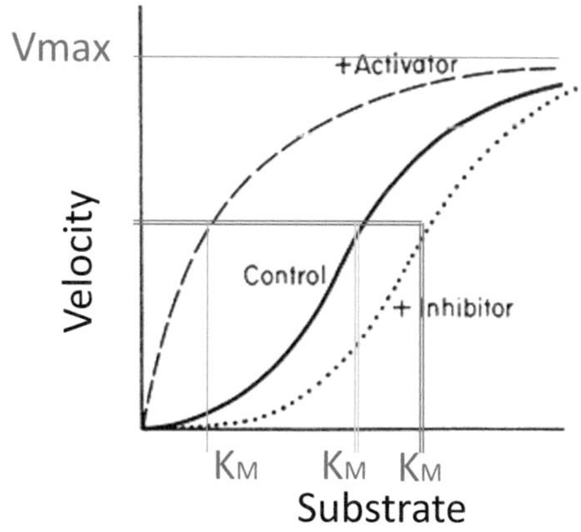

Fig. 9 Allosteric activators and inhibitors, which do not follow Michaelis Menten kinetics. At constant substrate concentration, the presence of allosteric activator leads to more or less constant value of V_{max}, but K_M changes. The value of K_M is decreasing (blue line) with respect to control (yellow line), whereas, if value for K_M is increasing, it will be allosteric inhibitor (red line)

4.5 Allosteric Regulation and Enzyme Kinetics

Michaelis–Menten equation allows us to work on rate of product formation with respect to substrate concentration, where the substrate will typically bind with active site of the enzyme. Few enzymes have an active site and also have a different site called allosteric site.

$$V_0 = \frac{V_{max}[S]}{K_M + [S]}$$

Allosteric sites are sites or places on the enzyme where any regulator can bind. These can be present anywhere on the enzyme and do not have fixed number. These allosteric sites can be influenced by binding of "allosteric activators" or "allosteric inhibitors." Allosteric activators increase the activity of the enzyme by activating them, while allosteric inhibitors decrease enzymatic reaction by deactivating them [1, 2, 5].

4.5.1 Allosteric Activators and Inhibitors

Now at constant substrate concentration, there are two ways to increase enzymatic activity or V_0, by increasing or decreasing K_M. Thus, in Michaelis–Menten equation (*see* Fig. 9), V_{max} value remains more or less equal, but K_M changes. If the value of K_M decreases, it is allosteric activator (Fig. 9 see activator and decrease in K_M value). If value for K_M increases, it will be allosteric inhibitor (Fig. 9 see inhibitor and increase in K_M value).

In Fig. 9b, regulator seems to change V_{max} significantly instead of K_M. Activator is increasing V_{max} at same substrate concentration while inhibitor is reducing V_{max} at same substrate concentration (Fig. 9b).

In the curve shown in Fig. 9b, regulator seems to change V_{max}, instead of K_M. Inhibitor decreases V_{max} significantly. Therefore, allosteric regulators can bind to allosteric sites on an enzyme, they can influence kinetics by increasing or decreasing K_M or V_{max}.

Feedback loop: *Consider a series reaction where A changes to E with intermediates B, C, D (and product E) using enzymes I, J, K and L respectively. Here product E, has effects on enzyme I, if it exerts positive effect on enzyme I activity, it is called positive feedback loop as molecule E increases the rate of reaction 1 and causes more E to be formed.*

$$A \underset{1(I)}{\rightarrow} B \underset{2(J)}{\rightarrow} C \underset{3(K)}{\rightarrow} D \underset{4(L)}{\rightarrow} E$$

If E has negative effect on enzyme I, this is negative feedback loop as molecule E decreases the rate of reaction 1 which ultimately leads to decrease in the rate of reaction involved in forming E. An example is phosphofructokinase (PFK) enzyme of glycolysis that uses ATP and catalyzes conversion of fructose-6-phosphate into Fructose 1,6 bisphosphate. ATP is a cosubstrate as well as an allosteric inhibitor of PFK. Thus, at higher ATP levels, cell says ATP is not required and no PFK activity is needed; therefore, it slows down the rate of ATP production. As ATP in this reaction is both a cosubstrate and an allosteric regulator, it is a homotropic inhibitor (substrate and regulator are the same molecules).

AMP (monomeric form of ATP) is the activator of PFK. If AMP levels are high, it indicates ATP levels are low; thus, cell requires ATP. AMP is a regulating molecule but not an active site substrate, it is called heterotropic activator (substrate and regulator are different molecules). Therefore, specific reactions make excellent control points for long multistep processes.

Usually the control steps are those that have highly negative ΔG, indicating that the step is not easily reversed, that is, it is an irreversible reaction. Feedback loops offer control of metabolic reactions with highly committing steps (with highly negative ΔG).

Allosteric enzymes do not follow Michaelis–Menten kinetics as shown in Fig. 9, the curve is sigmoid rather than hyperbola. Allosteric regulators associate with the enzyme at a precise regulatory site resulting in conformational or electrostatic alterations with concomitant increased or decreased enzyme activity. The enzymes that have the precise allosteric sites are referred to as allosteric enzymes. Molecules binding to allosteric site of enzymes and influencing binding on a different site are referred to as effectors.

Binding can be cooperative (binding of an effector at one site on the enzyme alters the binding affinity at another site) or if binding of an effector can reduce binding affinity for other ligands it is allosteric inhibition or negative allosteric modulation.

5 Notes

1. All the reagents should be at room temperature. Usage of ice-cold reagent should be avoided as they may not give the appropriate results [11].
2. Add the enzyme just before taking absorbance so that initial readings are not skipped [12].
3. In order to ensure that the sample contains the enzyme and appropriate results are being obtained, setup a reaction (1) without sample, (2) boil the sample and then perform the reaction, (3) use the sample and then record the absorbance at particular recommended wavelength. If the absorbance is noted in (1) and (2) case, that indicates false result. Only (3) case should give the desirable results.
4. If assay is not working, perform reaction with reagents at room temperature, check protocol and recheck pH of the buffer, use the recommended wavelength, perform assay in correct cuvette [11, 12].
5. In case inconsistent values are obtained, check the buffers along with its pH, check for sample degradation, read at an appropriate wavelength, or prepare samples in recommended buffer [11].
6. Sample degradation problems—Always aliquot samples, try to use aliquoted sample only once, that is, avoid freeze–thaw cycles, and check for interfering substances as detergents or inhibitors in the sample buffer [11, 12].
7. Inconsistent reading of sample and standard—Always use properly stored samples; aliquot and use. Thawing should be complete; if sample or buffer is not completely thawed, it would lead to errors. Avoid using expired or very old reagents [11, 12].
8. Pipetting errors should be avoided, appropriate dilutions should be followed, and experimental protocol should be checked.

References

1. Jeremy MB, John LT, Lubert S (2002) Biochemistry, 5th edn. W H Freeman, New York; ISBN-10: 0-7167-3051-0
2. Voet D, Voet JG, Pratt CW (2013) Fundamentals of biochemistry: life at the molecular level, 4th edn. Wiley, Hoboken, NJ; ISBN 13: 9780470570951
3. Michaelis L, Menten M (1913) Die kinetik der invertinwirkung. Biochem Z 49:333–369
4. Sakharov IY, Makarova IE, Ermolin GA (1988) Purification and characterization of intestinal alkaline phosphatase from harp seal. Comp Biochem Physiol B 90:709–714
5. Lehninger AL, Nelson DL, Cox MM (2000) Lehninger principles of biochemistry. Worth Publishers, New York
6. Kapojos J, Poelstra K, Borghuis T, Van den Berg A, Baelde H, Klok P, Bakker W (2003) Induction of glomerular alkaline phosphatase after challenge with lipopolysaccharide. Int J Exp Pathol 84:135–144
7. Poelstra K, Bakker W, Klok P, Hardonk M, Meijer D (1997) A physiologic function for alkaline phosphatase: endotoxin detoxification. Lab Investig 76:319–328
8. Lineweaver H, Burk D (1934) The determination of enzyme dissociation constants. J Am Chem Soc 56(3):658–666. https://doi.org/10.1021/ja01318a036
9. Eadie GS (1942) The inhibition of cholinesterase by Physostigmine and Prostigmine. J Biol Chem 146:85–93
10. Hofstee BHJ (1959) Non-inverted versus inverted plots in enzyme kinetics. Nature 184 (4695):1296–1298. https://doi.org/10.1038/1841296b0
11. http://docs.abcam.com/pdf/protocols/Troubleshooing-guide-for-enzymatic-assay-kits.pdf
12. Bisswanger H, (2014) Enzyme assays. Perspectives in Science 1(1–6):41–55

Chapter 15

A Novel Technique for the Detection of LncRNAs on Tissue Sections

Andrew E. Massey, Manish K. Tripathi, Murali M. Yallapu, Meena Jaggi, and Subhash C. Chauhan

Abstract

Long noncoding RNA (LncRNA) is a class of noncoding RNA that typically consists of over 200 nucleotides that do not encode proteins. It is thought that lncRNA accounts for most of the noncoding transcriptome in humans, as tens of thousands of lncRNA transcripts have been identified. However, only a few lncRNAs have been characterized in detail to date. It is known that they act as important regulators of gene expression and are thought to have multiple functions in cellular and developmental processes. In this chapter we will discuss about the technique for detecting lncRNA from tissue section.

Keywords LncRNAs, Therapeutic target, Z-probes

1 Introduction

Long non-coding RNAs (lncRNAs) have attracted interest for their role in different diseases, however considerable data needs to be unraveled regarding their exact mechanisms in these conditions. They are discovered and analyzed by different types of genomic technologies. Although there are different kinds of lncRNAs, little is known about their function and regulation. Recent transcriptomic technologies have been used to uncover the novel regulatory functions which lncRNAs may have in the regulation of different metabolic processes. The regulation of gene expression is a very important process in the cell. LncRNAs play an important role as they can act as both gene inhibitors and activators *via* various mechanisms, adding complexity to the overall understanding of genetic regulation [1, 2].

LncRNA molecules have been implicated in multiple disease states that are of significance in cancer [3–6]. Recent studies have implicated a role of lncRNA in pancreatic cancer [7], gastric cancer [8], breast cancer [9], and ovarian cancer [10], among various

Neelima Gupta and Varsha Gupta (eds.), *Experimental Protocols in Biotechnology*, Springer Protocols Handbooks,
https://doi.org/10.1007/978-1-0716-0607-0_15,

other cancers. This class of molecule can play a diverse set of roles in cancer, as it can have either a transcriptional or translational effect. In addition, various lncRNA molecules can have either protumorigenic or tumor suppressive properties: significant examples of protumorigenic lncRNAs include MALAT1, HULC, HOTAIR, and H19, while some tumor suppressive lncRNAs include MEG3 and GAS5. Given the disparate roles of these different lncRNAs, they can also act as therapeutic targets—novel technologies are being investigated to either enhance the actions of tumor suppressive lncRNA or downregulate oncogenic strands of lncRNA [11, 12].

2 Technique

Several different methods currently exist to detect and visualize lncRNA in biological samples, including a recently developed technique called RNA Scope®, a method for in situ hybridization (ISH) of RNA molecules. A visual summary of this procedure is shown in Fig. 1 [13]. This process involves the use of a complementary nucleic acid probe to localize or visualize a specific sequence within a tissue section. This separates it from the similar technique of immunohistochemistry, which is commonly used to visualize proteins within tissue. RNAscope® can be used with formalin-fixed, paraffin-embedded (FFPE) tissue samples, which can then be visualized under a brightfield microscope, or a fluorescent microscope if using fluorescent tags. This procedure allows for multiplex detection of up to four genes, although this number is based on the limitations of fluorescent dyes that can be discernible from one another spectrally. Since it can be used to detect gene expression in situ, it makes this highly sensitive technique a promising translational approach for various RNA biomarkers [13–15].

3 Materials

Formalin-fixed, paraffin embedded (FFPE) tissue samples.

Formalin solution, neutral buffered, 10% (Sigma-Aldrich, #HT501128) Superfrost Plus Slides (Fisher Scientific, #12-550-15).

Slide Warmer (Lab Scientific, #XH-2004).

Tissue-Tek Slide Staining Set (Andwin Scientific, #NC073-1403) Xylene (Fisher Scientific, #X3P-1GAL).

Reagent Alcohol, HistoPrep (Fischer Scientific, #HC-600) RNA-scope® Hydrogen Peroxide (ACDBio, #322335).

NxGen Decloaking Chamber System (BioCare Medical, #DC2012).

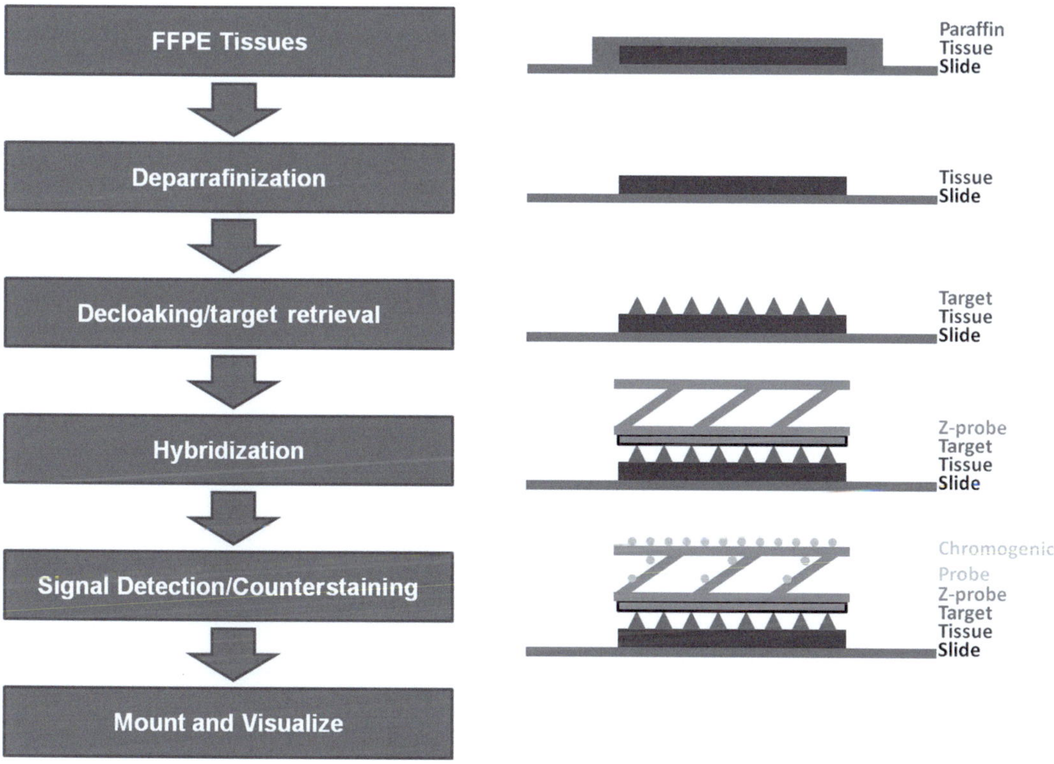

Fig. 1 Schematic overview of the RNA Scope® procedure. [13]

RNA Scope® Target Probes (e.g., RNA Scope® 2.5 VS Probe-HS-MUC13; ACDBio, #312919) RNA Scope® HD detection kit (RED) (ACDBio #322360).

RNA Scope® Target Retrieval Reagents (ACDBio, #322000) ImmEdge hydrophobic barrier pen (Vector Labs, #H-4000) RNA Scope® Protease Plus (ACDBio, #322331)

HybEZ Hybridization System (ACDBio, #310010) RNA Scope® Wash Buffer, 50× (ACDBio, #310091) Hematoxylin (Fisher Scientific, #H345-25).

Ammonium hydroxide, 28–30% solution in water (ACROS Organics, #AC423300?).

EcoMount (Biocare Medical, #EM897L).

24 mm × 50 mm glass coverslips (VWR, #48404-453).

4 Methods

4.1 RNA Scope® General Overview [13]

Step 1: Two Z probes hybridize to lncRNA transcript.

Each target probe (known as a Z probe) has 18–25 base complementary sequences to target and hybridize to lncRNA molecules of interest. These probes also contain 14 base sequences complementary to the preamplifier. These are bound sequentially on the surface of the lncRNA in pairs over a region of approximately 50 bases, forming a pair of Z probes ("double Z").

Step 2: Hybridization of the preamplifier to 28 complementary sequences formed from combination of two Z probes.

Preamplifiers are added and bind to "double ZZ" (a pair of Z probes bound to lncRNA, making a total of 28 bases). This region of 28 bases is then used to hybridize the preamplifiers, which in turn contain up to 20 binding sites for amplifiers.

Step 3: Amplifiers bind to multiple binding sites on the preamplifier.

Amplifiers are then bound to preamplifiers, giving a larger surface area for chromogenic probes to bind and allow for stronger fluorescence of targeted genetic material.

Step 4: Chromogenic labeled probes bind to multiple binding sites on each amplifier.

With addition of amplifiers, multiple probes can bind on each amplifier (20 probes per amplifier), intensifying signal that is detected. This allows for a considerable up regulation of the signal strength. For example, if visualizing a 2-kb region of an RNA molecule, 40 probe pairs would attach, leading to a theoretical total of up to 16,000 labeling molecules per RNA molecule.

Before starting this procedure, tissue sections of interest are fixed in formalin (10% neutral buffered formalin, 24 h) then treated with alcohol (70% EtOH, 24–48 h). After treatment, the tissues are embedded in paraffin, sectioned to a thickness of 5–7 μm and mounted on Superfrost slides.

4.2 Z-Probe Procedure [13, 16]

1. Place FFPE tissue slides on slide warmer set to 60 °C for 2 h.
2. Place the slides into 250 mL of xylene for 5 min, then transfer to a second container with fresh xylene.
3. Incubate the slides in a gradient of alcohol dilutions (100%, 95%, 90%, 80%, 70%, 50%) for 5 min each.
4. Place the slides into the hydrogen peroxide dish and treat for 10 min.
5. Rinse the slides in the dish containing distilled water.
6. While deparaffinizing, mix RNA Scope® 25 mL Target Retrieval Regent (10×) with 225 mL distilled water to make a 1× solution.
7. Set up the HybEZ oven. First, turn on and set to 40 °C Then, place a humidifying paper, soaked in distilled water, into the humidity control tray. Insert covered tray into the oven and close the door. Warm the tray for 30 min before use.

8. Place the slides into the 1× Target Retrieval Reagent, then place inside of decloaking chamber. Fill the chamber to the requisite level with distilled water.
9. Turn on the chamber and run for 15–25 min. Note that the exact time at this step depends on the tissue and target gene. Remove the slides and rinse in distilled water.
10. Use the ImmEdge hydrophobic barrier pen to create a barrier around the tissue section on each slide. Let the barrier dry completely (let sit 5–10 min at room temperature).
11. Place the dried slides onto the HybEZ slide rack and add ~5 drops of RNA Scope® Protease Plus to cover each tissue section.
12. Remove the humidity control tray form the HybEZ oven and insert the slide rack. Close the lid and incubate at 40 °C for 15–25 min. Again, the exact time will depend on the tissue sample.
13. Remove the tray and remove any excess liquid by gently tapping the slides. Place into a dish with distilled water (250 mL). Wash the slides in fresh water 3–5 times.
14. While incubating, prepare wash buffer by diluting 60 mL of 50× buffer into 2.94 L of distilled water. Also, prepare staining reagents: dilute hematoxylin to 50% in distilled water (total volume 200 mL). Prepare 0.02% ammonia water solution by adding 167 μL of 30% ammonium hydroxide solution into 250 mL distilled water under a fume hood.
15. Retrieve AMPs 1–6 from the 4 °C refrigerator and place at room temperature. Gently warm in a 40 °C water bath for 10 min just before use. Gently swirl the vials to mix.
16. To begin hybridization, gently remove any excess liquid on the slides and add ~4 drops of target probe to cover the tissue. Place tissues in the oven for 2 h at 40 °C.
17. Remove the slides and promptly remove excess liquid then place into 1× wash buffer.

    Wash for 2 min at room temperature two times, each in fresh 1× wash buffer.
18. To hybridize, use the following temperatures and times for each of the AMPs in order, with a rinse as described in **step 17** in between each incubation: AMP 1 (30 min, 40 °C); AMP 2 (14 min, 40 °C); AMP 3 (30 min, 40 °C); AMP 4 (15 min, 40 °C); AMP 5 (30 min, room temperature).
19. For signal detection, briefly centrifuge Fast RED-B tube to ensure contents are at the bottom of the tube before opening. Prepare enough RED detection solution to properly cover the tissues by mixing RED-A to RED-B in a 60:1 ratio (e.g., 3 μL of RED-B and 180 μL of RED-A).

20. Take each slide and remove excess liquid. Pipet ~100 μL of RED solution onto tissue sections. Ensure tissues are completely covered. Place the slides into humidity control tray and incubate for 10 min at room temperature.
21. Remove the slides and gently remove RED solution by tilting the slides over a waste container and tapping to remove excess staining solution. Promptly place the slides into Tissue-Tek dish filled with distilled water and rinse for 2 min. Replace with fresh water and rinse again.
22. To counterstain the slides, place them into a Tissue-Tek dish containing 50% hematoxylin solution and stain for 2 min.
23. Promptly place into staining dish with tap water and gently agitate the slides up and down to rinse. Repeat with fresh water until the slides are clear but the tissue sections are purple.
24. Replace water with 0.02% ammonia water solution. Gently agitate the slides up and down 2–3 times. The tissue should turn blue.
25. Replace with tap water and gently rinse the slides 3–5 times.
26. Remove the slides from the staining dish and dry the slides completely, then briefly dip the slides into a fresh dish containing 250 mL xylene. Place 1–2 drops of EcoMount on the slides before xylene dries. Carefully place a 24 × 50 mm coverslip over the tissue section, avoiding air bubbles. Repeat for any remaining slides, then let air-dry. Tissues can then be visualized for RNA staining.

5 Notes

1. The majority of lncRNAs possess low conservation of sequences compared to protein coding genes; therefore, searching for their homologs in other related species is difficult. This necessitates requirement of sequence determination specific to each lncRNA for the respective organism [1].
2. Largely annotated data for lncRNA for the target organism of interest can be found on genome browsers including Ensembl (www.ensembl.org) and UCSC (www.genome.ucsc.edu). Other computational and/or experimental data can be found on other websites (e.g., http://www.lncipedia.org) [17].
3. LncRNA molecules have a naturally low abundance, and they can also base-pair with mRNA to trigger its degradation; this can be determined by a simple basic local alignment search tool [18].

References

1. Huarte M (2015) The emerging role of lncRNAs in cancer. Nat Med 21 (11):1253–1261. https://doi.org/10.1038/nm.3981
2. Ferrè F, Colantoni A, Helmer-Citterich M (2016) Revealing protein-lncRNA interaction. Brief Bioinform 17(1):106–116. https://doi.org/10.1093/bib/bbv031
3. Schmitz SU, Grote P, Herrmann BG (2016) Mechanisms of long noncoding RNA function in development and disease. Cell Mol Life Sci 73(13):2491–2509. https://doi.org/10.1007/s00018-016-2174-5
4. Poller W, Dimmeler S, Heymans S et al (2018) Non-coding RNAs in cardiovascular diseases: diagnostic and therapeutic perspectives. Eur Heart J 39(29):2704–2716. https://doi.org/10.1093/eurheartj/ehx165
5. Wang P (2018) The opening of Pandora's box: an emerging role of long noncoding RNA in viral infections. Front Immunol 9:3138. https://doi.org/10.3389/fimmu.2018.03138
6. Sun M, Kraus WL (2015) From discovery to function: the expanding roles of long noncoding RNAs in physiology and disease. Endocr Rev 36(1):25–64. https://doi.org/10.1210/er.2014-1034
7. Lv Y, Huang S (2019) Role of non-coding RNA in pancreatic cancer. Oncol Lett 18 (4):3963–3973. https://doi.org/10.3892/ol.2019.10758
8. Fattahi S, Kosari-Monfared M, Golpour M et al (2019) LncRNAs as potential diagnostic and prognostic biomarkers in gastric cancer: a novel approach to personalized medicine. J Cell Physiol 235(4):3189–3206. https://doi.org/10.1002/jcp.29260
9. Youness RA, Gad MZ (2019) Long non-coding RNAs: functional regulatory players in breast cancer. Noncoding RNA Res 4(1):36–44. https://doi.org/10.1016/j.ncrna.2019.01.003
10. Tripathi MK, Doxtater K, Keramatnia F et al (2018) Role of lncRNAs in ovarian cancer: defining new biomarkers for therapeutic purposes. Drug Discov Today 23(9):1635–1643. https://doi.org/10.1016/j.drudis.2018.04.010
11. Parsons C, Tayoun AM, Benado BD et al (2018) The role of long noncoding RNAs in cancer metastasis. J Cancer Metastasis Treat 4 (4):19. https://doi.org/10.20517/2394-4722.2018.11
12. Jiang M-C, Ni J-J, Cui W-Y, Wang B-Y, Zhuo W (2019) Emerging roles of lncRNA in cancer and therapeutic opportunities. Am J Cancer Res 9(7):1354–1366. http://www.ncbi.nlm.nih.gov/pubmed/31392074.
13. Tripathi MK, Zacheaus C, Doxtater K et al (2018) Z probe, an efficient tool for characterizing long non-coding RNA in FFPE tissues. Noncoding RNA 4(3):1–10. https://doi.org/10.3390/ncrna4030020
14. Amini Chermahini G, Rashnonejad A, Harper SQ (2019) RNAscope in situ hybridization-based method for detecting DUX4 RNA expression in vitro. RNA 25(9):1211–1217. https://doi.org/10.1261/rna.070177.118
15. Xie F, Timme KA, Wood JR (2018) Using single molecule mRNA fluorescent in situ hybridization (RNA-FISH) to quantify mRNAs in individual murine oocytes and embryos. Sci Rep 8(1):1–11. https://doi.org/10.1038/s41598-018-26345-0
16. Wang F, Flanagan J, Su N et al (2012) RNAscope: a novel in situ RNA analysis platform for formalin-fixed, paraffin-embedded tissues. J Mol Diagn 14(1):22–29. https://doi.org/10.1016/j.jmoldx.2011.08.002
17. Rühle F, Stoll M (2016) Long non-coding RNA databases in cardiovascular research. Genomics Proteomics Bioinformatics 14:191–199
18. Freedman JE, Miano JM (2017) Challenges and opportunities in linking long noncoding RNAs to cardiovascular, lung, and blood diseases. Arterioscler Thromb Vasc Biol 37:21–25

Index

A

B

C

D

E

F

G

H

I

L

M

Neelima Gupta and Varsha Gupta (eds.), *Experimental Protocols in Biotechnology*, Springer Protocols Handbooks, https://doi.org/10.1007/978-1-0716-0607-0,

MIX
Papier aus verantwortungsvollen Quellen
Paper from responsible sources
FSC® C105338

If you have any concerns about our products, you can contact us on
ProductSafety@springernature.com

In case Publisher is established outside the EU, the EU authorized representative is:
Springer Nature Customer Service Center GmbH
Europaplatz 3, 69115 Heidelberg, Germany

Printed by Libri Plureos GmbH
in Hamburg, Germany